PPL REVISION 1200 QUESTIONS AND ANSWERS FOR THE PRIVATE PILOT

R.D.CAMPBELL AND R.J.HALL

BSP PROFESSIONAL BOOKS

OXFORD LONDON EDINBURGH

BOSTON MELBOURNE

First published in Great Britain by
    Granada Publishing 1982
Reprinted 1983 (ISBN 0 246 11882 2)
Reprinted by Collins Professional and
    Technical Books 1986
Reprinted with updates by
    BSP Professional Books 1988
Reprinted with minor updates 1990

British Library
Cataloguing in Publication Data
Campbell, R.D. (Ronald D.)
    P.P.L. revision.
    1. Aeroplanes, Flying – Questions &
    answers – For private pilots
    I. Title   II. Hall, R.J.
    629.123′5217′076

ISBN 0–632–02402–X

BSP Professional Books
A division of Blackwell Scientific
    Publications Ltd
Editorial Offices:
Osney Mead, Oxford OX2 0EL
    (Orders: Tel. 0865 240201)
25 John Street, London WC1N 2BL
23 Ainslie Place, Edinburgh EH3 6AJ
3 Cambridge Center, Suite 208,
    Cambridge, MA 02142, USA
54 University Street, Carlton,
    Victoria 3053, Australia

Printed and bound in Great Britain
at the Alden Press, Oxford

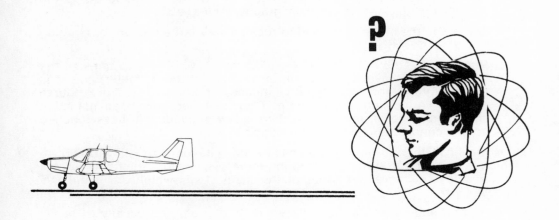

INTRODUCTION

One of the problems which arise during learning and afterwards lies in the difficulty of retaining the knowledge acquired. During a properly constructed Private Pilot's Course the student will be required to absorb many thousands of 'learning items' and to illustrate this point one has only to consider the volume of knowledge which has to be recalled and used, during the few seconds in which a pilot obtains a 'taxy clearance'.

In considering the number of items which have had to be learned in order to accomplish this minor task let us dwell for a moment on the very limited aviation knowledge which individuals may have when they first commence their flying training. They will most probably never have seen an aircraft radio, or used a headset or microphone. They will not know what runway numerals signify, or how to recognize an altimeter or understand the terminology QFE or QNH, etc.

However by the time a person undertakes early solo flights the few seconds spent in obtaining a taxy clearance will be crammed with hundreds of learning items gained from the first day onwards of his flying course. Just how many items he needs to know in order to talk in an intelligent way to an Air Traffic Controller can be seen by the various mental and physical actions which will be necessary. For example:

He will need to recognize a radio set, know how to switch it on and select the correct frequency. For a pilot to learn (as against being told) which frequency to use he will first have to know of the existence of the UK Air Pilot or alternative aeronautical publications, he will have to know where in the publication the information is listed, and how to interpret such terms as "Tower, Approach, etc.", and what the ATC hours of operation are for a particular aerodrome.

The pilot now needs to know what to say to ATC and how to interpret the reply he is given. This will immediately involve him in the use of the Phonetic Alphabet (26 learning items plus numerals and other phrases). Most ATC taxy clearance calls will need to include information regarding the aircraft's endurance and the purpose of the flight. The student will, therefore, need to know how to identify and interpret the fuel gauges and then to apply this information by mental calculation to give the aircraft endurance in relation to the intended flight. To be able to do this he will have had to study the aircraft manual and interpolate graphs/tables, etc.

At this stage the student has had to recall a very large number of learning items, but to go on . . . .

Runway 26, QFE 1009 mb, QNH 1003 mb ? . . . . just to think of the learning items involved in an understanding of the altimeter, how it is set and interpreted and what it all means in relation to ATC procedures and the various phases of flight. The point has now been made! A short exchange on the radio lasting a few seconds will have touched upon a considerable sum of knowledge.

However in this particular example we have taken a repetitive exercise which is carried out during every flight, therefore memory retention will not be difficult to achieve, but what of the many items which are used only occasionally ?

Any knowledge which is used infrequently can be, and usually is, easily forgotten and there are thousands of items which come within this category when operating an aeroplane. Items which suddenly have to be recalled and which the pilot has forgotten! And although there will be occasions when the pilot will be able to find the answer in reference material there will also be many occasions when this type of reference material is not readily available and the pilot may find himself in the position of flying an aircraft whilst lacking the essential information for doing so in safety.

It is for this reason that this book has been produced. It will prove of value to any pilot who takes the time and trouble to set himself the task of answering a number of questions at intervals of time and by so doing strengthening his ability to recall knowledge as and when required to do so, and in some cases to gain new knowledge. In summary, tackling the following statements and checking your answers will:

Focus attention, provoke curiosity, and stimulate interest.

Assist the recall of known facts.

Introduce new but essential knowledge.

Provoke the pilot to raise further questions.

Encourage the pilot to revise and/or research answers.

The questions which follow are all statements of the TRUE/FALSE type, this method is one which with its many variations in question content has a wide range of usage. It is well adapted to the testing of knowledge and facts especially when there are only two possible answers. It does, however, lead to the tendency for the reader to guess the answer. Therefore, self discipline will be required to obviate this and bring the greatest benefit to the person doing the test. If the answer is not known it will be more beneficial for the reader to go back to the appropriate reference material and re-establish the knowledge required to answer the question with confidence.

A point to bear in mind when answering the questions is that some have a mixture of correct and incorrect information, thus making the statement partially true. This is done quite deliberately in order to test whether the reader has sufficient knowledge of the subject.

The book is divided into two sections, questions 1 to 600 are contained in the first section and in the second section these 600 questions are restated but often with variations, this is done to re-inforce memory retention. We should also like to point out that if question No. 1 is true then question No. 601 will be false and so on throughout. All answers are shown at the rear of the book.

Although every care has been used in compiling the questions and answers, it must be appreciated that aviation is not a static thing and in consequence Air Legislation, Regulations, Air Traffic Rules and Flight Procedures are always subject to change. This should be borne in mind, as some of the questions and answers may become outdated at any time after the publication date of the book.

Reference Material

The list of publications which contain the information from which the statements were devised, and to which the reader may need to refer where an incorrect answer is chosen or doubt exists in deciding whether the statement is true or false are as follows:

The Air Navigation Order.

The Air Navigation (General) Regulations.

UK Air Pilot and NOTAMS.

Aeronautical Information Circulars.

1:500,000 Aeronautical Chart ICAO.

1:250,000 Topographical Air Chart of the United Kingdom.

British Civil Airworthiness Requirements.

Flying Training for the Private Pilot (AOPA).

Student Manual Part 2.

Ground Training Manual No. 1. Air Legislation, Aviation Law, Flight Rules and Procedures. Air Traffic Rules and Services.

Ground Training Manual No. 2. Air Navigation and Aviation Meteorology.

Ground Training Manual No. 3. Principles of Flight, Airframes and Aero Engines, Aircraft Airworthiness, Aircraft Instruments.

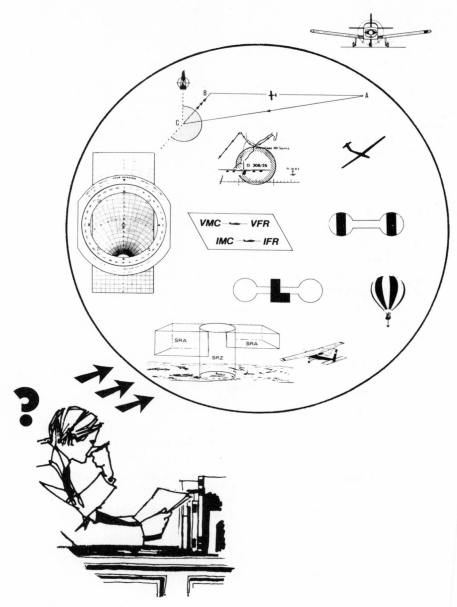

How good is your recall ? . . . .

It's on the tip of my tongue

It'll come back to me in a minute

My mind's gone completely blank

I'll need time to think that one out

Of course I know, well I think I do . . . .

Now you can find out — just turn the page and start.

Section   1

1. The maximum lift/drag coefficient is achieved by adjusting the angle of attack.

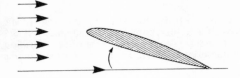

2. (MET) The adjacent symbol indicates FOG.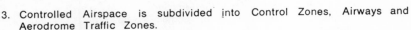

3. Controlled Airspace is subdivided into Control Zones, Airways and Aerodrome Traffic Zones.

4. The term **CAVOK** means that visibility is 10 nm or above, that there is no cloud below 2,000 ft and no cumulonimbus, precipitation, thunderstorms, shallow fog or low drifting snow.

5. Once notice of ETA has been sent to the destination aerodrome, the commander of an aircraft must inform the destination authorities, as quickly as possible, of any estimated delay of 2 hours or more.

6. A Certificate of Release to Service must be issued by a qualified aircraft engineer whenever any equipment essential to the airworthiness of the aircraft is overhauled, repaired, replaced or modified.

7.  A double arrow on one side of a vector triangle of velocities is used to indicate TRACK and GROUND SPEED.

8. XXX In Morse Code, transmitted by sound (except RTF) or visual signalling indicates that the commander of the aircraft has an urgent message to transmit.

9. (SIGNAL AREA) Black numerals on a yellow background indicates wind speed.

10. TAF's are Aerodrome Weather Reports.

11.  If the engine temperature is high and the oil pressure is normal the pilot should enrich the mixture, decrease the throttle setting and check that the cowl flaps are open.

12. The UK is divided into a number of Altimeter Setting Regions (ASR's) in which an area QNH is forecast for every hour.

13. If an aircraft is loaded so that the C of G is towards the aft limit, its stability will be slightly reduced but the shortened moment arm will increase the effectiveness of the tail surfaces.

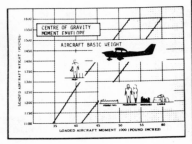

14. Once awarded, the PPL is non-expiring but the privileges it contains will lapse unless the holder completes a minimum of 5 hours flight time as pilot in command (or equivalent experience) certified in the holder's log book by an authorised Private Pilot Licence examiner every 13 months.

15. During a descending turn, the airspeed will tend to increase.

16. 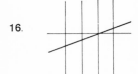 In relation to charts 'conformality' is not a requirement for bearings or angles to be presented accurately.

17. The atmospheric pressure rises as a warm front approaches.

18. (SIGNALS AREA) A single red letter "L" indicates that light aircraft are permitted to take-off and land either on a runway or on the area designated by a white letter "L".

19. The Equator is both a great circle and a Rhumb line.

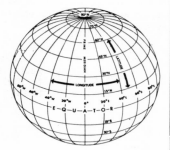

20. 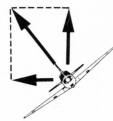 Small arms may be carried aboard light aircraft provided that they are not loaded and that the ammunition is stowed separately at a distance from any possible fire hazard.

21. Warm air can contain more water vapour than cold air.

22. (MET) The abbreviation Cs means: cloud surface.

23. In uncontrolled airspace, the holder of a PPL may fly at an altitude of over 3,000 ft amsl only provided there are no passengers on board.

24. A 30° angle of bank increases the stalling speed by less than 10%.

25. If the pressure and temperature of a given volume of air remains constant, its density will vary in inverse proportion to the relative humidity.

26. A steady red light signal from aerodrome control to an aeroplane in flight means: DO NOT LAND, THE AERODROME IS NOT AVAILABLE FOR LANDING.

27. For the first few thousand feet amsl, the atmospheric pressure drops by approximately 1mb per 100 ft.

28. The airspeed for the maximum angle of climb is about 10 knots less than the airspeed for the maximum rate of climb.

29. A flashing white light signal from aerodrome control to an aeroplane on the ground means RETURN TO THE STARTING POINT ON THE AERODROME.

30. (MET) The adjacent symbol indicates 'Wind westerly, backing.'

31. When determining right-of-way between converging aircraft, an aircraft towing a glider does not have to give way to gliders.

32. On the "Half Million" chart, contourlines mark every 500 ft amsl. The layer tinting shows white for 0 ft to 500 ft, ochre from 500 ft to 1,000 ft and changes to a deeper colour for each 1,000 ft thereafter.

33. The ground speed of an aircraft is calculated by converting the TAS in knots into statute miles per hour.

34. (MET) The abbreviation **FG** signifies "FOG".

35. Frise ailerons are designed to increase the drag on the raised wing.

36. (MET) A Col is an area with a strong pressure gradient.

37. The number appearing at the top of a C of A is the code number of the flight manual which comprises part of the certificate.

4235

38. Vfe is the maximum safe speed with flaps extended.

39.  If the external static source were to become blocked during flight and if there were no alternate source inside the cabin, the Altimeter, Vertical Speed Indicator and the Airspeed Indicator readings would all return to zero.

40. Mecator projections are not normally used for pilot navigation except in equatorial latitudes.

41. The curvature of the surface of an aerofoil is called the "camber".

42. (MET) On a significant weather chart the abbreviation **FRQ** means : 'Frost at this altitude'.

43. One advantage of the 1:250.000 Topographical chart is that it shows the base of Airways.

44. The term "Cloud Ceiling" denotes the altitude amsl above which there is less than one eighth cloud.

45. In Temperate Regions, high cloud formations (Cirrus, Cirrostratus and Cirrocumulus) occur mainly between 16,500 ft and 45,000 ft.

46.  (SIGNALS AREA) A red and yellow striped arrow placed along two adjacent sides of the signals area and pointing in a clockwise direction indicates aircraft turning right after take-off.

47. In Britain when an aeroplane is flying towards an area of higher pressure, the prevailing wind will be from the left and its altimeter will over-read.

48. Overheating and detonation in aero engines is avoided by operating on a rich fuel/air mixture.

49. In the visual signal code used by survivors, the adjacent symbol means "Am proceeding in this direction".

50. (MAP) ➤ This symbol indicates a Hang Gliding Site.

51. Stratus clouds begin to form in unstable atmospheric conditions.

52. Information relating to Customs facilities at aerodromes can be found in the AGA Section of the UK Air Pilot.

53. Aerodrome Control Units control all aerodrome traffic not being handled by Approach Control Units, VFR flights in the aerodrome circuit, and all ground manoeuvres.

54. "In Command" flight time refers to flights when you are the only pilot operating the controls.

55. In the Northern Hemisphere a person standing with his back to the wind will have the high pressure area on his right.

56. The "Air Navigation Order" is the only publication through which the CAA promulgates its orders.

57. (MET) ᴗ On a Significant Weather Chart this symbol indicates Turbulence.

58. Aircraft placed in the Special Category may be certificated for Aerial Work where appropriate.

59. A "Rate 3 turn" is one in which the aircraft turns 360° in one third of a minute, i.e. 20 seconds.

60. Distress and Diversion cells of the RAF maintain a continuous watch on 121.5 MHz.

61. The Centre of Gravity arm of an aircraft is equal to the total moment divided by the total all up weight.

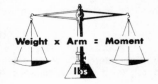

62. Radiation Fog is formed when a warm, moist sea breeze blows gently over the colder, neighbouring land surface.

63. When flying in the vicinity of an active thunderstorm allow one mile clearance for every 200 ft of its vertical development.

64. In order to comply with the 'Quadrantal Rule' a Flight Level of 170 should be used by aeroplanes cruising on a magnetic track of 170°.

65. (MET) The abbreviation Ns means: Nimbostratus.

66. Detailed information about airfields in the UK is to be found in the AGA Section, Vol. 2 of the "UK Air Pilot".

67. Detailed information on areas and locations which are restricted for reasons of safety is to be found in the RAC Section, Vol. 1 of the "UK Air Pilot".

68. Water vapour weighs more than dry air.

69. (MARSHALLING SIGNAL) Left arm down; right arm repeatedly moved upward and backward, means: OPEN UP PORT ENGINE or TURN TO STARBOARD.

70. All air navigation obstructions more than 500 ft agl are lighted and any unserviceability of the lighting is promulgated in a Class I NOTAM. Obstructions between 300 ft and 500 ft agl may or may not be lighted, and unserviceability of any lighting will not normally be promulgated by NOTAM.

71.  Pilots who have obtained the permission of the ATCU concerned to enter a Special Rules Zone or Area, and who have notified the ATCU of their position, level and track must maintain a continuous watch on the notified radio frequency as long as they are within the SRA or SRZ.

72. In the visual signal code used by survivors, the adjacent symbol means "No assistance is required". N

73. The "One in Sixty Method" of assessing track error involves multiplying the distance flown by the distance off track and dividing the result by 60.

74. Runway 08 is in use. The surface wind is 020/20 so the crosswind component is 17 kts.

75. (MET) This symbol indicates Cold Front.

76. True track minus westerly variation equals magnetic track.

77. The visual signal code symbol for "Yes" is shown on the right. Y

78. With increase of altitude between ground level and 2000 ft winds tend to veer.

79. An aircraft may be flown between the Vno and the Vne in any meteorological conditions provided that the pilot does not use abrupt and extreme control movements.

80. Aircraft, Engine and Variable Pitch Propeller Log Books must be kept up-to-date in respect of flight times and preserved by the operator/owner of the aircraft for two years after the aircraft/engine/propeller has been destroyed or withdrawn from use

81. You are approaching from the south and you report airfield in sight. You are instructed "G-RN JOIN LEFT BASE RUNWAY TWO TWO". In zero wind conditions your 'base leg' heading should be 130°.

82. Military aerodromes notified as available for civil use are listed in AGA 2 in the same manner as civil aerodromes, but such aerodromes may not be used (except in diversion or emergency situations), without obtaining permission prior to take-off.

83. Control Zones are established around major airports from ground level to 3,000 ft aal.

84. When a pilot who has filed a Flight Plan diverts, or lands at any aerodrome not specified in his Flight Plan, he is required to inform his destination aerodrome within 24 hours.

85. (MET) The adjacent symbol indicates "Showers".

86. Within the United Kingdom an aircraft shall not carry out instrument approach practice when flying in Visual Meteorological Conditions.

87. Military Low Level Routes where high speed flying takes place are marked on the "Half Million Chart" by lines of black dashes.

88. A free balloon is not an aircraft.

89. Over the UK the change in the magnetic variation is relatively slow, averaging about 0.l of a degree annually.

90. Clear ice may form on the leading edges of aeroplanes flying through clouds of ice particles e.g. Cirrus.

91. At night, projectiles showing bursting, red or green lights or stars, in the vicinity of an aircraft indicate that a prohibited area is being overflown.

92. (MARSHALLING SIGNAL) Raise arm and hand, with fingers extended horizontally across the body, then clench fist, means : ENGAGE BRAKES.

93. When the wet bulb thermometer registers almost the same temperature as the dry bulb thermometer, the humidity is low.

94. When flying from an area of higher temperature to an area of lower temperature, the altimeter will under-read.

95. If the owner of an aeroplane in the Special Category chooses to maintain his Aircraft in accordance with an approved maintenance schedule, the validity of the C of A may be increased from 1 to 2 years.

96. By day, surface winds usually tend to back and decrease in strength.

97. "High Density Altitude" is used to describe a situation where the temperature is abnormally high and the density is abnormally low at a particular altitude.

98. If you increase your touch-down speed by 5% you will increase your ground roll by 5%.

99. The use of flap increases both lift and drag.

100. (MET) The abbreviation SN means "Snow".

101. ⊔⊔ (MET) The adjacent symbol on a Significant Weather Chart signifies, 'Heavy rain'.

102. Single seater aircraft with a maximum authorised weight not exceeding 910 kg, in the Special Category, may be certificated to permit hiring from one person to another.

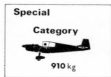

Special Category

910 kg

103. All nominated Customs and Excise Airports within the UK provide Customs facilities on a 24 hour basis.

104. It needs only a small amount of frost and/or ice on the wings and control surfaces to endanger take-off performance.

105. Topographical Air Charts of the United Kingdom (1:250,000) are overprinted with Aeronautical information up to a limiit of 5,000 ft amsl, but not above.

106. Apart from certain sections within TMA's and CZ's, Airways are under permanent IFR.

107.  Cumulus clouds begin to form in stable atmospheric conditions.

108. Magnetic track minus westerly variation equals true track.

109. (MET) On the Significant Weather Chart the abreviation **CLD** means 'Cold'.

110. Advection Fog is formed when warmer air is cooled to the dew point by passing over a colder surface.

111. A helicopter is an aeroplane.

112. The commander of a flying machine on an aerodrome shall not be required to obey Marshalling Signals if, in his opinion, it is inadvisable to do so in the interests of safety.

113. All aircraft operating under IFR in Controlled Airspace are required to conform to the Quadrantal Rule.

114. Bird sanctuaries are marked on topographical maps and they normally extend from the surface to 4,000 ft agl.

115. Information about use of airfields—fees and charges; the Public Health, Customs and Immigration regulations; the import and export of goods—are all found in the FAL Section of Vol. 2 of the "UK Air Pilot".

116. 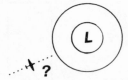 When flying towards an area of deteriorating weather, you can expect the aircraft to drift to the left.

117. (MET) The adjacent symbol signifies 'High cloud only'.

118. (MET) The abreviation **OVC** on a Significant Weather Chart means: 'Overcast'.

119. If the outside air temperature is ten degrees lower than the ISA for that altitude, the altimeter will over-read by about 5%.

120. Identification Beacons flash a two-letter morse group every 12 seconds: green for civil aerodromes and red for military aerodromes.

121. When rising above 'ground effect', the aircraft's nose will tend to pitch downwards.

122. Except for areas. of actual precipitation, visibility when flying in warm frontal conditions is better than visibility when flying in cold frontal conditions.

123. Anti-Collision Lights are coloured ground lights used to mark out the manoeuvring area.

124. If the external static source were to become blocked and an alternate static source inside the cabin were to be used, the Airspeed Indicator and the Altimeter would both over-read.

125. (MET) The abbreviation **CAT** on a Significant Weather Chart indicates 'Clear Air Turbulence'.

126. The lower end of the white arc shown on an Air Speed Indicator dial indicates the minimum speed at which the flap can be lowered.

127. The thickness of a wing at any point is called its "Camber".

128. If the C of G of an aircraft is moved aft, it is more likely, if mishandled, to go into a spin and, once in a spin, it will be more difficult to recover.

129. If the weight of an aircraft is increased the wing loading will be increased.

130. The various conditions, to which the C of A is subject, are typed in detail on the front page of the certificate.

131. A triple arrow on one side of a vector triangle of velocities is used to indicate TRACK and GROUND SPEED.

132. Thunderstorms may move in any direction, even opposite to that of the prevailing wind.

133. Small training aircraft are designed to encounter gust velocities of at least 100 ft per second.

134.  All aircraft must be weighed before a Certificate of Airworthiness can be issued, determination of the centre of gravity is only necessary for aircraft weighing more than 910 kg.

135. (MET) This symbol ▲▲▲ indicates 'Cold front'.

136. "Cross Country Flight" is defined in the ANO as a flight of not less than 50 nm with a landing at destination.

137. A red pyrotechnic signal from aerodrome control to an aeroplane in flight means: DO NOT LAND, WAIT FOR PERMISSION.

138. A "Rate 4 Turn" is one in which the aircraft turns 360° in 4 minutes.

139. **Airways are normally 10 nm wide and extend vertically from a specific** altitude to an upper limit which is usually about 25,000 feet amsl.

140. Aero piston engines develop less power when the humidity of the air is high.

141. (MET) The abbreviation **FR** indicates 'Freezing'.

142. Wind speed and direction at the "surface" are measured at a standard height of 6 ft above the ground.

143. The Dew Point may be defined as either 0° C or 32° F.
144. Compass deviation alters with longitude and a compass deviation card is fitted in the cockpit.

145. Spinning is classified as an aerobatic maneouvre.

146. (MET) The abbreviation **Sc** indicates: "Scattered cloud".

147. Small flap settings increase lift more than drag.

148. The "Air Navigation (General) Regulations" is a Civil Air Publication which gives details of general regulations relating to light aircraft.

149. (MET) **GRADU** means : 'Slowly lifting'.

150. In good weather when air traffic movements are light, permission for Special VFR flight in a control zone may be given to pilots who hold neither the IMC nor the Instrument Rating provided that the visibility is not less than 5 nm.

151. A light aircraft taking-off from an airfield situated 1,000 ft amsl when the temperature is 25°C will require about 25% more in the length of the take-off run.

152. The Lambert Conformal Conic Projection is only used in polar regions.

153. When two aircraft are approaching each other approximately head-on, each shall, whether on the ground or in the air, alter course to the right.

154. Details of all Royal Flights will be promulgated by special RF NOTAM and all controlled airspace used for this purpose, whether permanent or temporary, will be notified as being IFR.

155. (MET) On a Significant Weather Chart the adjacent symbol indicates : 'Severe airframe icing'.

156. When flying beneath a TMA but outside the Control Zone, pilots must set the altimeter datum to the QFE of any aerodrome within the Control Zone.

157. The various scales of instrumentation and equipment considered essential to match differing circumstances of operation (by day or night, over land or over water etc.) are left to the discretion of the operator/owner's experience and the manufacturer's recommendations.

158. When by night an intermittent white luminous beam is directed at an aircraft, it means that the aircraft is in danger of overflying a Restricted or Prohibited area.

159. In uncontrolled airspace at less than 3,000 ft amsl the holder of a PPL may not carry passengers unless he can remain clear of cloud and in sight of the surface with a flight visibility of not less than 5 nm.

160.  The adjacent altimeter is indicating: 8,880 feet.

161. You are approaching from the north-west and you report airfield in sight. You are instructed "G-RN JOIN RIGHT BASE RUNWAY TWO SIX". Your base leg heading should therefore be 170 degrees.

162. (MET) The adjacent symbol indicates : 'Overcast'.

163. Impact icing can affect engines using float type carburettors or fuel injection systems.

164. The term "Night" means the time between half an hour before sunset and half an hour after dawn.

165. The wind veers after the passage of a warm front and backs after the passage of a cold front.

166. When an anti-collision light is fitted it shall be displayed when an aircraft is on the apron, by day or night with its engine(s) running.

167.  (MARSHALLING SIGNAL) A circular motion of the right hand at head level with the left arm pointing to the appropriate engine means: 'START ENGINE(s)'.

168. The aerofoil sections of stabilators and tailplanes are usually asymmetrical.

169. When an aircraft has drifted off track, the "Closing Angle" method of calculating the Heading Alteration Required involves dividing the Estimated Closing Angle by the Proportion of Track Flown.

170. (MARSHALLING SIGNAL) Arms extended, palms facing inwards, then swung from the extended position inwards, means : 'CHOCKS INSERTED'.

171. (SIGNALS AREA) A checkered flag or board containing 12 equal squares, coloured red and yellow alternately, signifies that the airfield is under the control of an ATCU.

172. In "AIRMET" forecasts the levels of clouds are given as height above ground level.

173. An increase in air density will produce more lift and more drag.

174. A Lower Airspace Radar Advisory Service is available to all aircraft in uncontrolled airspace within 80 nm of each participating airfield when flying between 3.000 ft amsl and FL 245.

175. The commander of an aircraft, on request by a competent authority, must produce all aircraft documentation and flight crew licences within 3 working days.

176.  There is virtually no coriolis effect above 2,000 ft agl and winds above that level blow unhindered across the isobars directly from high pressure to low pressure.

177. **The "Minimum Safety Altitude" is generally accepted as being 1,000 ft** above the highest ground/obstruction within 5 to 10 nm either side of the appropriate segment of the aircraft's track and within 5 to 10 nm radius of the point of departure and the destination.

178. A 60° angle of bank doubles the stalling speed.

179. The AGA section of the UK Air Pilot lists all TORA, TODA, ED and LDA distances in feet.

180. In the code designation of Restricted, Danger and Prohibited Areas, the numbers after the oblique show the altitude in thousands of feet amsl below which the particular area may not be overflown.

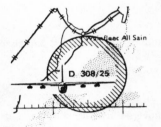

181. **The forecast QNH is the lowest expected reading for the area during** the period of validity.

182. A meridian is half a great circle and is a rhumb line.

183. An aircraft shall not fly lower than 500 feet above any person vessel, vehicle or structure.

184. (SIGNALS AREA) A double white cross and/or two red balls suspended from a mast one above the other means that repairs are in progress and that the runways are temporarily unserviceable.

185. The category in which an aircraft may be operated is shown on the C of A.

186. The "True Track" of an aircraft is its actual path over the ground measured from True North.

187. (MARSHALLING SIGNAL) Right or left arm down; other arm moved across the body and extended to indicate position of the other marshaller, means: TURN AIRCRAFT AS INDICATED and TAKE UP POSITION IN FRONT OF NEW MARSHALLER.

188. A flashing green light signal from aerodrome control to an aeroplane in flight means: 'CLEARED TO LAND'.

189. By night, surface winds usually tend to back and decrease in strength.

190. If the oil pressure gauge reads Zero and the oil temperature gauge and/or cylinder head temperature gauge reads normal, the pilot should make an emergency landing immediately.

191. A pilot who wishes to operate in a Special Rules Zone, must obtain ATC clearance.

192. If the temperature is 15°C at sea level, it will be approximately 5°C at 5,000 ft and −5°C at 10,000 ft.

193. 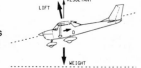 (MET) The adjacent symbol indicates: 'Cumulus'.

194. The glide ratio of an aircraft is equal to its lift/weight ratio.

195. The UK is covered by only 2 FIR's, but the FIR's for each region are divided up geographically between different ATCC's using the same call sign but different frequencies.

196. **Va** is the maximum safe speed in turbulent air.

197. In a spin, the airspeed remains low but the pilot's first reaction should be to close the throttle.

198. Runway 08 is in use The surface wind is 060/20 so the crosswind component is 15 kts.

199. In straight and level flight, in calm air, an aeroplane is subjected to a load factor of +I.

200.  For the purpose of renewal of the privileges of the PPL, the equivalent of 5 hours flight time as pilot in command may be obtained by completing at least 3 hours in command plus sufficient dual flying with a qualified instructor to make up a total of 5 hours.

201. Prior to the issue of a Private Pilot's Licence, the CAA require the following minimum of flying training:-
     (a) Not less than 10 hours dual.
     (b) Not less than 10 hours solo.
     (c) A total of not less than 40 hours.

202. Before making a flight from any aerodrome with an ATCU, you must either "Book Out" or file a flight plan.

203. There are no variations in the portrayal of angles and bearings on a Lambert Conformal Chart, but because the area is distorted, woods and other shapes depicted will not have the same shape as the actual ground features.

204.  A licenced pilot who operates and/or owns an aircraft (not used for public transport and with a maximum permitted weight of 2730 kg or below) can find the list of maintenance tasks he is legally empowered to carry out in the "Air Navigation (General) Regulations".

205. If the Relative Humidity is 30%, this means that 30% of the atmosphere, by volume, is composed of water vapour.

206. Two or more white crosses displayed on a runway or taxiway with their arms at 45° to the centre line indicate that the section marked by them is unfit for the movement of aircraft.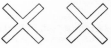

207. The term "Rolling Resistance" refers to an aircraft's lateral stability along the longitudinal axis.

208. Below 2,000 ft the wind will follow the direction of the isobars more closely over the sea than over the land.

209. In a low pressure area, the air is decending.

210. (MET) The adjacent symbol indicates: "Freezing Rain".

211. A flashing green light signal from aerodrome control to an aeroplane on the ground means: "CLEARED FOR TAKE-OFF".

212. ICAO Aeronautical Charts of the United Kingdom (1:500,000) are valid for the calendar year in which they are issued (1st Jan.—31st Dec.)

213. An aircraft shall not fly over or within 3,000 ft of any assembly in the open air of more than 1,000 persons witnessing or participating in an organised event; (without the written permission of the CAA and organisers of the event) or below such a height as would enable it to alight clear of the assembly in the event of engine failure.

214. (MARSHALLING SIGNAL) Right arm raised at the elbow, with the arm facing forward, means : „ALL CLEAR, MARSHALLING FINNISHED".

215. A glider, when flying at night, may either display the same lights as are mandatory for aeroplanes of 5,700 kg or less, or a single red light showing in all directions.

216. The wings of light aircraft are normally designed so that the outer section will stall before the inner section.

217. Aircraft in Special Category Class I will be permitted to operate in the Public Transport Category.

218. The Meteorological Office will issue a Gale Warning when winds in excess of 33 kts and/or gusts in excess of 42 kts are expected near the surface.

219. The aspect ratio of a wing is calculated by dividing the span by the mean chord.

220. A parallel of latitude is not a great circle but it is a rhumb line.

221. Runway 36 is in use. The surface wind is 030/24 so the crosswind component is 8 kts.

222. Most of the Range and Endurance tables/graphs shown in aircraft manuals are calculated on the basis that the mixture control will be used in the 'Rich' position.

223. True track plus easterly variation equals magnetic track.

224. Terminal Control Areas are established around the upper altitudes of most Control Zones where Airways intersect.

225. The Basic Weight consists of the empty weight of the aircraft but includes the weight of unusable fuel and oil together with fixed items of equipment as are listed on the weight schedule.

226. An Area Forecast is normally issued once per day at 0600 hours.

227. On a significant Weather Chart dashed lines are used to show areas of similar weather conditions.

228. The entry procedures for specific Control Zones are promulgated in the RAC Section of the UK Air Pilot.

229. An aircraft voltage regulator controls the rate of charge to the battery from the alternator/generator.

230. Private Pilots who hold an "Instrument Rating" are permitted to enter Control Zones, but only after Special VFR Flight clearance has been obtained.

231. The presence of stratus clouds indicates turbulent flying conditions.

232. The use of "Carburettor Heat" increases fuel consumption by up to 20%.

233. Unlike the UK Air Pilot, many aircraft manuals still list the various take-off and landing distances in feet.

234. In relation to Danger Areas, "day" means from 0800 to 1800 GMT.

235. The flying instruction in light aircraft which is needed for a person to receive a licence may be conducted from unlicenced aerodromes.

236. Information on aerodromes and private landing strips is to be found in the AGA Section of the UK Air Pilot.

237.  The adjacent altimeter indicates : 8,800 feet.

238. Below 3,000 ft amsl, in uncontrolled airspace and with passengers, the holder of a PPL (without an IMC Rating), must remain in a flight visibility of at least 3 nm.

239. Survivors of a forced landing, on seeing a SAR plane's green pyrotechnic, should wait for a full 15 seconds before firing the first red (if available) answering pyrotechnic.

240. The angle of incidence of the tailplane is usually slightly greater than that of the wings in order to support the weight of the tail unit.

241. When the holder of a PPL has not completed the necessary flight time within the statutory 13 monts he will automatically be required to complete 10 hours dual as well as 5 hours solo before his privileges are renewed.

242. (MET) The abbreviation **LYR** on a Significant Weather Chart means : "Last reported observation".

243 When rising above ground effect, the aircraft's acceleration will suddenly increase.

244. By night, a Sea Breeze, often blows from the sea to the land.

245. The airfield you are approaching has three runways; 04/22, 08/26 and 18/36. You are instructed "G-RN JOIN LEFT BASE RUNWAY 04." Your correct base leg heading will be approximately 130 degrees.

246. The temperature and the dew point will fall after the passage of a warm front.

247. Limiting load factors are measured for both positive and negative "g".

248.  All civil aerodromes have Aerodrome Traffic Zones with a horizontal radius of 1.5 nm from the centre of the aerodrome.

249. Frontal Fog is caused by the cooling effect of cold air as it rises over a warm front.

250. During the hours of operation of an aerodrome Air Traffic Unit a pilot must comply with the directions issued by Air Traffic Control unless he considers it unsafe to do so.

251. Prohibited Areas need only be avoided during the hours of daylight.

252. High temperature and high humidity increase both lift and thrust during take-off.

253. The ICAO World Aeronautical Chart Series (1:1,000,000) has too small a scale to be interpreted easily during flight.

254. MARSHALLING SIGNAL) Either arm and hand placed level with chest, then moved laterally with the palm downwards, means: "CUT ENGINES"

255. When light aircraft are used in the Public Transport Category the servicing periods are calculated in both calendar months and flying hours: one month or 50 hours being a typical figure (which ever is the sooner).

256. When the Dry Bulb temperature is close to the Wet Bulb temperature and light winds prevail, conditions will be conducive to clearing skies and good visibility

257. Under no circumstances may the holder of a PPL without an "Instrument Rating" fly across the base of an Airway without prior clearance from the appropriate ATCU.

258.   (SIGNALS AREA) A red square panel with a yellow strip along each diagonal signifies that the airfield is unsafe for the movement of aircraft and that landing on the airfield is prohibited.

259. The "Manoueuvring Area" includes all parts of the aerodrome except those areas set aside for the take-off and landing of aircraft.

260. When flying towards an area of deteriorating weather conditions, your altimeter will begin to over-read.

261. The standard Transition Altitude for civil aerodromes outside Controlled Airspace is 3,000 ft QNH.

262. An aeroplane placed in the Transport Category may be operated for the purpose of transport only.

263. The "Rules of the Air & Traffic Control Regulations" is a document containing law, enacted by Parliament.

264. Calibrated Airspeed is the American equivalent of Indicated Airspeed.

265. The aircraft Flight Manual does not form part of the Certificate of Airworthiness.

Flight/Owner's Manual.

Pilot's Operating Handbook.

266. The term "drift" means the angular difference between the track and heading required to maintain the track.

267. (MET) The abbreviation **SCT** means: "Scattered".

268. (MET) The adjacent symbol signifies a "Severe squall line".

269. Low density air produces more lift and less drag.

270. A climb from sea level to 3,000 ft will result in a larger drop in pressure in cold air than in warm air.

271. Magneto's are used to provide an electric current which creates an electric spark from the sparking plugs in each cylinder.

272. Icing of the external engine air intake does not present a hazard as the rapid flow of air prevents the formation of ice except in real blizzard conditions.

273. (MET) The abbreviation **TS** means: "Thunderstorm".

274. An overtaking aircraft shall, whether on the ground or in the air, alter course to starboard untill well clear of the aircraft being overtaken.

275. In the absence of Identification Beacons, Aerodrome Beacons flash a blue light during conditions of poor visibility.

276. A level turn at a bank angle of 30° increases the load factor to less than 1.2.

277. Large cumulus clouds often develop vertically more rapidly than the maximum rate of climb of light aircraft.

278. If, by mistake, a pilot sets the QNH on his Altimeter 10 millibars higher than its actual value, he will be under the impression that his altitude is about 300 ft lower than it really is.

279. If the C of G of an aircraft is moved forward, the aircraft's stability in pitch decreases.

280.  (MARSHALLING SIGNAL) Arms down, the palms facing outwards, then swung outwards, means: "CHOCKS AWAY".

281. A pilot who wishes to operate in a Special Rules Zone or Area must first obtain ATC clearance.

282. The higher the atmospheric pressure at ground level, the greater the lift developed by the wings during take-off.

283. The maps and charts used in pilot navigation use Magnetic North as a reference datum.

284. (MET) The abbreviation **AC** means: "Altitude of cloud".

285. A red pyrotechnic light or flare signal from an aircraft to aerodrome control means: "I AM COMPELLED TO LAND".

286. The International Standard Atmosphere serves as an arbitrary datum to measure aircraft performance and calibration of aircraft instruments.

287. "Notified" is a term used to indicate the pilot's intentions to land or go round again.

288. When the Carburettor Hot Air system is in use the induction air will automatically pass through a filter to remove dust and other particles from the air ingested by the engine.

289. **Vno** is the normal operating speed.

290. An airship is an aircraft.

291. (MET) The term Tempo means: "Temporarily".

292. Lift increases in proportion to the airspeed, but drag increases in proportion to the square of the airspeed.

293. The carriage of radio is mandatory for all flights conducted over water.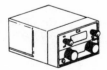

294. The adjacent symbol shown on a Significant Weather Chart means: "Forecast wind speed and direction".   ⟶ 25

295. (SIGNAL AREA) A white "T" indicates that aeroplanes and gliders shall take-off and land in the direction that is parallel with the shaft, and towards the crossarm, of the "T".

296. The Attitude Indicator (Artificial Horizon) is operated by an external venturi or through the static/pressure system.

297. Information relating to Nationality and Registration Marks of aircraft, Civil Aviation Legislation and Air Navigation Regulations are all to be found in the GEN Section. Vol 2 of the "UK Air Pilot".

298. When the altimeter datum is set at QNH, the reading will show the altitude amsl.

299. 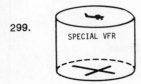 Pilots holding an IMC Rating may be granted Special VFR Flight permission in Control Zones provided visibility is not less than 3 nm and air traffic conditions are suitable.

300. High Intensity Radio Transmission Areas are not shown on 1:500,000 topographical charts.

301. RAC Section, Vol.1, of the UK Air Pilot lists all customs airports which can be used for entry to, or departure from, the UK.

302. (SIGNALS AREA) A red square panel with a yellow diagonal strip signifies that the state of the manoeuvring area is poor and that pilots must exercise special care when landing.

303. Buys Ballot's Law:- If you stand with your back to the wind in the northern hemisphere, the low pressure area will be on your right.

304. An aeroplane must never be flown unless a C of A has been issued.

305. The basic weight of an aircraft is the weight of the aircraft, its basic equipment and its fuel and oil, but it does not include the weight of the pilot, passengers and their baggage.

306. (MET) **OCNL** means: "Occasional".

307. An aeroplane will glide further with the flaps in the raised position.

308. A level turn at a bank angle of 60° increases the load factor to 2.

309. The colour of AVGAS 100 LL is green.

310. An aeroplane placed in the Private Category must have its C of A renewed every 3 years.

311. It is the operator's responsibility to ensure that the Certificate of Airworthiness is valid prior to the pilot taking off.

312. No cloud formation can take place in air which is not saturated i.e. when the temperature is above the Dew Point.

313.  Indicated Airspeed, when corrected for instrument and position error, is known as Rectified Airspeed.

314. The average Lapse Rate is about 2°C per 1,000 ft increase in altitude.

315. "Track error" is the angular difference between the track required and the actual track made good.

316. Modern light aircraft are so constructed that a pilot would black out long before the airframe could be subjeced to excessive load factors.

317. The filing of a Flight Plan is not required before flights conducted under **IFR** within Controlled Airspace.

318. The adjacent Aeronautical Chart Symbol means: ⋁⋀⋁⋀⋁⋀⋁
     Outer Boundary of TMA or CTA.

319. (MET) On a significant Weather Chart the adjacent symbol means: "Moderate Turbulence".   ⎯⋀⎯

320. When a quantity of air is compressed into a smaller volume it becomes cooler.

321.  When the altimeter datum is set at QFE, the reading will show the height aal.

322. The operation of the aircraft outside the conditions in the C of A will not invalidate the C of A but will render both the aircraft's and pilot's insurance policies void.

323. If the wind is from dead ahead on take-off it will normally be from several degrees to the right at 2,000 ft agl.

324. There is no legislation applicable to the carriage of emergency equipment for Private Flights.

325. Permanent (Scheduled) Danger Areas are shown on the "Chart of UK Airspace Restrictions" with a broken red outline.

326. The sublimation of hoar frost on the exterior surfaces of an aircraft occurs only when the aircraft is parked in sub-zero temperatures.

327. If the holder of a PPL has not flown as pilot in command for 26 months since the expiry of his privileges he will have to undertake 2 hours ground instruction and 12 hours flying instruction.

328. An aeroplane shall not fly over any congested area of a city, town or settlement:
(i) Below such a height as would enable it to be flown clear of the area and land without incurring danger to persons or property in the event of engine failure; or
(ii) Below a height of 1500 ft above the highest fixed object within 2,000 ft of the aircraft whichever is the higher.

329. Provided the "Relative Humidity" is high, carburettor icing may take place when the temperature of the outside air is between +20° and −60° C.

330. (SIGNALS AREA) A white disc displayed above the crossarm of a white "T" and/or a black ball suspended from a mast signifies winds gusting to over 35 kts.

331. Short grass on firm ground will increase the take-off run of a light aircraft by approximately 20%.

332. The CAP 53 contains information about licensing and Student and Private Pilot privileges.

333. Recovery from a spin depends first on controlling the yaw.

334.    (MARSHALLING SIGNAL) Raise arm, with fist clenched, horizontally across front of body, then extend fingers, means: OPEN THROTTLE.

335. All aerodromes in the UK, both licenced and unlicenced, are required to exhibit either an Identification Beacon or an Aerodrome Beacon H24 every day.

336. The commander of an aircraft is responsible for furnishing an Accident Report if any accident occurs causing injury to a person who is on the aircraft.

337. The regulations concerning 'Certificates of Test and Experience' in relation to pilot licensing, are contained in the Air Navigation Order.

338. The "Order" means the current ANO (as amended).

339. The difference between True and Magnetic heading is known as "Magnetic Variation".

340. The adjacent Aeronautical Chart symbol means: ⊷ ⊶ ⊶   "High Tension Wires".

341. Glider sites may have winching cables extending to 3,000 ft above the surface.

342. The scale of radio equipment installed in each individual aircraft is found in the aircraft's normal documentation in the form of a Certificate of Approval of an Aircraft Radio Installation.

343. Runway 36 is in use. The surface wind is 350/30. Therefore the crosswind component is 5 kts.

344. **Vne** is the never exceed speed.

345. Certain military aerodromes available for civil use have a Transition Altitude above 3,000 ft and pilots using these aerodromes must use QNE below this altitude.

346. (MET) The adjacent symbol means:  "Fog decreasing".

347. Pilots must file a Flight Plan for a Special VFR Flight with the appropriate Air Traffic Control Unit.

348. The reception range of VHF transmissions depends entirely upon the strength of the transmitter.

349. (MET) The adjacent symbol signifies: "Cumulonimbus".

350. A Lambert Conformal Conic Projection produces great circles which are, for all practical purposes, straight lines.

351. The Air Traffic Advisory Service directs all flights under IFR in Airways, TMA's and Control Zones.

352. The "Mass Balance" method of reducing the possibility of control surface flutter involves the addition of weights so placed as to move the C of G of the control surface forward of its hinge line.

353. Stratus clouds may cover a very large area but are never more than 1500 feet in vertical extent.

354. The density of the air is directly proportional to the atmospheric pressure.

355. A Special VFR Flight clearance may be granted to a Private Pilot who is unable to comply with IFR, provided traffic and weather conditions are suitable.

356. On Lambert Conformal Conic Charts all meridians are vertical.

357.  A white cross displayed at the end of a runway indicates that it is the upwind end of the runway in use.

358. (MET) The abbreviation Inter means: "Intermittent".

359. Magnetic track minus easterly variation equals true track.

360. Wings with a low aspect ratio have a higher stalling angle than wings with a high aspect ratio.

361. At aerodromes not having an ATCU on watch, and at which Take-offs and Landings are not confined to a runway, a flying machine or glider when landing shall leave clear on its left any aircraft which has already landed, is already landing or is about to take off.

362. An aircraft signalling by white flashes, switching on and off its landing and/or navigation lights means: IMMEDIATE ASSISTANCE IS REQUIRED.

363. A steady green light signal from Aerodrome Control to an aeroplane in flight means: CLEARED TO LAND.

364. By day, a Land Breeze normally blows from the land to the sea during the afternoon.

365. Private aircraft required by the owner for private flying only may be placed in the Special Category.

366. Strong updraughts and down draughts are often met just below large cumulus clouds or near the surface.

367. Meteorological Warnings, e.g. Gales, Frost, Snow, etc., are automatically issued to all aerodromes.

368. Without the permission of the aircraft's owner/operator, no person may enter an aircraft without a police warrant.

369. If the Alternator/Generator fails during flight, the operation of the magnetos will not be affected.

370.  As a general rule, the Centre of Pressure moves forward when the angle of attack is increased.

371. In a trough of low pressure, the air is ascending expanding and cooling.

372. (MET) The adjacent symbol is used to indicate "Rain".   ●

373. The minimum age for the issue of a Private Pilot's Licence or the award of Student Pilot's privileges is 18 years.

374. The ICAO World Aeronautical Chart Series (1:1,000,000) is based on the Mercator Projection.

375. On a Mercator chart, areas become increasingly distorted towards the Equator.

376. The limits of acceptable error for an altimeter's indications are +30 ft and −50 ft.

377. If the carburettor ices up repeatedly, changing the RPM may solve the problem.

378. Outside controlled Airspace, certain Advisory Areas and Routes have been established and a specific ATC service is available on request.

379.  Rules of the Air and Air Traffic Services are to be found in the AGA Section in Volume 1 of the "UK Air Pilot".

380. (MAP) The adjacent symbol indicates :"Gliding is the primary activity at this location". Ⓖ

381. The corrections of current chart errors are promulgated in UK Air Pilot Amendment Sheets, and errors of operational significance are corrected immediately in NOTAMS or AICs.

382. The aspect ratio of a wing may be defined as span squared, divided by the area.

383. When flying away from an area of deteriorating weather conditions, you expect your altimeter to begin to over-read.

384. The temperature and the dew point rise when the cold front passes, but the pressure falls rapidly.

385. Carburettor Icing only occurs when flying in cloud or rain with outside air temperatures below 0°C.

386. A headwind component of 20% of the lift-off speed reduces the ground roll by 35%.

387. Warm humid air at ground level will cause an increased take-off run.

388. By telephoning the "AIRMET" Service and giving at least 2 hours' notice, a pilot can obtain a pre-recorded route forecast for any route within the UK not exceeding 500 nm.

389. The accuracy of the Heading Indicator is constant in straight and level flight; errors are caused only by friction on the gimbals resulting from manoeuvres.

390. A flashing red light signal from aerodrome control to an aeroplane in flight means: DO NOT LAND, THE AERODROME IS NOT AVAILABLE FOR LANDING.

391. On aeronautical charts, the adjacent symbol indicates a Special Access Lane Entry/Exit Point.

392. (MET) On Significant Weather Charts the adjacent symbol indicates: "Severe Line Squall".

393. Pilots of aeroplanes who wish to cross airways by day in VMC must obtain clearance via the FIR Service prior to take-off.

394. There is no legal requirement for the pilot to consider the likely weather en route or at destination if he intends to operate in accordance with the Visual Flight Rules.

395. When "QNH" is set on an altimeter the term "Altitude" should be used.

396. In uncontrolled airspace, not Notified to the contrary, whilst under IFR, at above 3,000 ft amsl, the aircraft must not be flown at a height of less than 1,500 ft above any obstruction within 2,000 ft of its track.

397.  A free balloon while flying at night shall display a steady red light, showing in all directions, suspended between 5 and 10 metres beneath its lowest part.

398. An aircraft, while landing or on a final approach to land shall have the right-of-way over other aircraft in flight or on the ground or water.

399. In so far as Customs requirements are concerned, flights to the Channel Islands and the Isle of Man are not considered as flights abroad.

400. (MARSHALLING SIGNAL) Arms placed above the head in a vertical position, means: THIS BAY OR AREA FOR PARKING.

401. Any line which crosses all meridians at a constant angle is called a "great circle" line.

402. (MET) Dry air has a higher density than moist air.

403. Control surfaces with horn balances are less sensitive than those without, and allow for firmer manual control by the pilot.

404. A flashing red light signal from aerodrome control to an aeroplane on the ground means: "STOP".

405.  When using non-inertial seat harnesses the shoulder harness should be tightened before the lower torso straps.

406. When flying in high pressure, anticyclonic conditions, one normally experiences clear skies and only light winds.

407. In relation to the term "Controlled VFR Flight" this procedure is not applicable within the UK.

408. A "Rate 1 Turn" is one in which the aircraft turns a complete circle, of 360 degrees, in one minute.

409. An aircraft may not be operated contrary to the conditions contained in its C of A unless the Pilot in Command at the time it is being so operated is the Sole Owner.

410. In the United Kingdom, by day or night, a series of projectiles discharged from the ground at intervals of 10 seconds, each showing on bursting, red and green lights or stars, shall indicate to an aircraft that it is flying in or about to enter a danger area or an area concerning national defence or affecting public interest and is required to leave the area or change course to avoid the area.

411. The pre-recorded forecasts of the "AIRMET" Service are updated twice daily at 0600Z and 1800Z.

412. The variable load of an aircraft is the weight of the crew, passengers and baggage.

413. Katabatic winds are caused by the cooling of air in contact with the surface in high mountainous areas and its flowing down to lower levels.

414. The angle of attack is the angle between the relative airflow and the mean chord line.

415. Roll clouds form on the leeward sides of mountains.

416. Clearance from the appropriate ATCU to make a special VFR Flight in a Control Zone does not absolve the pilot from complying with IFR.

417. 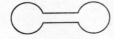 (SIGNALS AREA) A white dumbell signifies that movement of aircraft and gliders on the ground shall be confined to grass surfaces only.

418. All aircraft with fixed radio installations require radio licences.

419. Whenever an aircraft is to be flown with a different weight of load or distribution of load, a weight and balance calculation must be made.

420. Wings with a low aspect ratio will cause less vortex drag than wings with a high aspect ratio.

421. When two or more aircraft are approaching a place to land the lowest aircraft shall have right-of-way unless the ATCU has given priority to another aircraft (always excepting emergency situations).

422. Private Pilots must always make their requests for Special VFR clearance to enter or transit a Control Zone at least 10 minutes before take-off.

423. **Vso** is the stalling speed, no power, flaps up and a load factor of zero.

424. On pilot navigation maps, all lines joining places of equal magnetic variation are called Isogonals''.

425. (MARSHALLING SIGNAL) Arms repeatedly moved upward and backward, beckoning on-ward, means: "MOVE AHEAD."

426. The expansion space at the top of fuel tanks is to allow for the expansion of the fuel in "full" tanks when the plane takes off and climbs to a higher altitude.

427. **530** ∧
(323) /.
Air Navigation Obstructions reaching 300 feet or more amsl are listed in the AGA Section, Vol 2 of the "UK Air Pilot".

428. True Airspeed is Rectified Airspeed corrected for density error.

429. (MET) The adjacent symbol indicates, "Drizzle". ❜

430. (MET) **CU** means: "Cumulonimbus".

431. **Vra** is a specifically recommended airspeed for operating the aircraft at all up weight.

432. A steady green light signal from aerodrome control to an aeroplane on the ground means: "CLEARED FOR TAKE-OFF".

433. Regulations covering the operation of Commercial Aircraft are to be found in the COM Section in Vol 2 of the "UK Air Pilot".

434. **SIGMET** is a continuous VHF broadcast reporting on the actual weather at selected aerodromes.

435. On a Mercator chart, all rhumb lines except parallels and meridians will be curved.

436. The density of the air is inversely proportional to its Absolute Temperature.

437. Water droplets or ice crystals which reduce visibility at the Earth's surface to between 1,000 and 2.000 m are called "fog".

438. The Prime Meridian (Zero Degrees) passes through the Greenwich Observatory in London.

439. When flying away from an area of deteriorating weather conditions you can expect a left drift.

440. The two FIRs covering the UK and the surrounding waters are the London FIR and the Scottish FIR.

441. (MARSHALLING SIGNAL) Arms placed down with palms towards the ground, then moved up and down several times, means: "FLAPS DOWN".

442. Outside Controlled Airspace in ADR's an advisory service offers a continuous separation service between all aircraft in the same vicinity.

443.  When the Standard Setting is used, the altimeter will read the altitude (in feet) amsl.

444. On routine flights in the UK (with no aerobatics) a drift error on the Heading Indicator of 5 degrees or more is unacceptable.

445. Flights within notified SRZ do not require VFR clearance provided the aircraft remains clear of cloud and in sight of the surface.

446. Experimental Aircraft of all kinds are certificated in the Special Category or with a Permit to Fly.

447. After an outward bound aircraft has been cleared by Customs it may not land again in the UK except to take on persons and goods previously entered in form XS 29A.

448. Contours, spot heights and hill shading form the basis of depicting relief on UK topographical charts.

449. An aeroplane placed in the Private Category may be used for any purpose other than Public Transport or Aerial Work.

450. As a general rule, the higher pressure area will be to the left when you are taking off into the wind in the Northern Hemisphere.

451. Captive balloons and kites, flying above 60 metres at night, shall display groups of red and white lights: a triangle of flashing lights on the ground around the mooring point and pairs of red-over-white lights at specified distances up the cable.

452. (MAP) The adjacent symbol indicates an "NDB".

453. Long grass on firm ground will increase the take-off run by at least 25%, but soft ground may make take-off impossible.

454.  As a general rule, the Centre of Pressure moves rearwards when the angle of attack is decreased.

455. A white letter "L" on the manoeuvring area indicates that part which shall be used only for the taking-off and landing of light aircraft.

456. In Temperate Regions, medium cloud (Altostratus and Altocumulus) is found chiefly between 3,000 ft and 6,000 ft.

457. (MET) The abbreviation **CAST** signifies: "Overcast".

458. As a general rule, when flying above a layer of mist or haze, a pilot should not descend below the minimum safety altitude in order to reduce the slant angle when trying to recognise features on the surface.

459. When descending from 2,000 ft agl to ground level, a pilot should anticipate that the wind will veer.

460. The Airspeed Indicator is activated by a combination of pressures from both the pitot and static lines, but the Vertical Speed Indicator is operated by the static line only.

461. A Statute Mile equals 5280 feet and a Nautical Mile equals 6080 feet.

462.  The ground beneath the approach path is lower than the landing runway. In moderate to strong winds a pilot should expect significant updraughts to occur during the final stages of the approach.

463. AIS centres operate on a 24 hour basis.

464. Light aircraft engines develop maximum power in warm, low density air.

465. A pilot is required to file a Flight Plan if he wishes to take advantage of an Air Traffic Advisory Service.

466. A tailwind component of 20% of the lift-off speed increases the ground roll. by 45%.

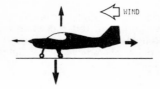

467. (MET) The abbreviation **FU** signifies: "SMOKE".

468. Meteorological units require 6 hours notice before issuing a Significant Weather Chart for a Route Forecast of less than 500 nm and 12 hours notice for a Route Forecast covering more than 500 nm.

469. A Class 3 medical certificate is required to qualify for the privileges of a Student Pilot or a Private Pilot's Licence and this will last for a period which will be determined by the age of the pilot.

470. (MET) The adjacent symbol signifies: "Hail".   ✳ ✳

471. The Volmet frequency in the Scottish FIR is different from that used in the London FIR.

472. In the UK, Temporary Danger Areas set up to cover areas where a disaster has occurred are not applicable to civil aircraft.

473. Notified Danger Areas are marked with a pecked red outline in the "Chart of UK Airspace Restrictions" and details of their limits are to be found in the ANO.

474. Cumulus clouds are sometimes embedded in large formations of stratus.

475. When a quantity of air expands into a larger volume it becomes warmer.

476. Special Rule Zones are only established around those aerodromes which are not protected by Control Zones.

477. CAP 85 is a booklet containing information on Aviation Law e.g. Rules of the Air and Air Traffic Control Regulations.

478. An aircraft in grave or imminent danger requesting immediate assistance should make one, or a combination of more than one, of the following signals:-
(a) By RTF the spoken word MAY DAY.
(b) By visual signalling, SOS in morse code.
(c) By a sucession of single red pyrotechnic lights fired at short intervals.
(d) By dropping a parachute flare showing a single red light.
(e) By sound signalling SOS in Morse Code (except on RTF).

479. An aircraft shall not fly closer than 500 ft to any person, vessel, animal, vehicle or structure.

480. The density of a given volume of air varies in direct proportion to its pressure.

481. In Temperate Regions, Cumulonimbus can form from near the surface to more than 40,000 ft.

482. (MET) The symbol on a Significant Weather Chart indicates: "Severe Icing".

483. The point on the airframe at which the wing root protrudes from the fuselage is called the "separation point".

484. (MET) The symbol  indicates: "Fog, Sky Visible".

485. 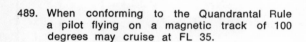 Fire extinguishers containing Carbon Tetrachloride are harmless to use.

486. At airspeeds up to approximately 100 knots the correct angle of bank for a Rate 1 turn is calculated by adding 5 knots to 10% of the indicated airspeed.

487. (MET) The letters RA in a TAF indicate: Rough Air Below 5000 ft.

488. When the temperature of water droplets fall below 0 degrees C the water particles will automatically turn into ice particles.

489. When conforming to the Quandrantal Rule a pilot flying on a magnetic track of 100 degrees may cruise at FL 35.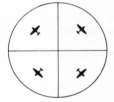

490. Pilots who fly fairly frequently should not offer to become blood donors.

491. (MET) The symbol indicates: "Spreading Stratocumulus".

492. Holders of Private Pilot Licences who have not flown as pilot in command for 4 years must apply to the Licensing Section of the CAA giving particulars of licences and ratings held, hours flown, etc. The CAA will then specify a training course to be undertaken before re-examination.

493. A white cross and a single white bar displayed at the end of a runway at a disused airfield indicates that the runway is unfit even for emergency use.

494. FAL Section, Vol 2 of the "UK Air Pilot" covers the facilities, runways, local flying restrictions, etc., of the Customs airports.

495. Thunderstorms present a particular hazard during an occlusion as the widespread stratus may mask the build-up of scattered cumulonimbus.

496. The term "Ground Visibility" refers to maximum height aal from which the surface can be seen at the time of reporting.

497. The specifications of the International Standard Atmosphere at mean sea level include an atmospheric pressure of 1013.25 and a temperature of 15 degrees C.

498. The topographical Air Charts of the United Kingdom (1:250,000) are constructed on a Transverse Mercator Projection.

499. On a Lambert Conformal Conic Projection, rhumb lines, with the exception of meridians, will appear as curved lines concave towards the nearer pole.

500. The lower density of the atmosphere above 10,000 ft significantly reduces the oxygen per unit volume. However, supplemental oxygen will not be needed unless flying above 20,000 ft.

501. When flying from lower pressure to higher pressure the altimeter over-reads.

502. The sequence of cirrus, cirrostratus, altostratus and then nimbostratus, heralds the passage of a warm front.

503. (MAP) The adjacent symbol denotes a DME:

504. During a Search and Rescue operation within the hours of daylight the search aircraft will flash its landing lights to signify the ground signal(s) has been understood.

505. The International Datum Level is used by all aircraft on the Continent of Europe while taking-off and landing.

506. The procedure to adopt in the event of an induction fire during engine starting is standard for all light aircraft.

507. Unless the pilot has prior permission from the ATCU an aircraft may not enter the aerodrome traffic zone in a Control Zone when the visibility is less than 5nm or the cloud ceiling less than 1,500 ft.

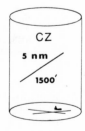

508. Military Air Traffic Zones vary in relation to their horizontal and vertical dimensions.

509. Before undertaking a flight which will necessitate entering a Control Zone, the pilot should consult the COM Section, Vol. 2 of the UK Air Pilot to ascertain the exact procedure to follow in case of radio failure.

510. Electrical fires in an aircraft cabin will normally be preceded by fumes and possibly smoke.

511. There is a danger of granular rime ice forming on the exterior of an aircraft descending through warm moist air.

512. You are approaching from the east and you report airfield in sight. You are instructed "G-RN JOIN DOWNWIND LEFT HAND CIRCUIT FOR RUNWAY THREE SIX" so you should descend to circuit height and turn right.

513. The RPM of a fixed pitch propeller at any given throttle setting will vary with the airspeed.

514. (MET) The abbreviation GR signifies Hail.

515. A heavily laden aircraft will have a higher lift off speed than when lightly laden.

516. The Aeronautical Emergency Service frequency 125.1 MHz should be used for Distress communications.

517. (MET) On a significant weather chart the letters CB are used to indicate 'Cloud Base'.

518.  A pilot who considers that his aircraft may have been endangered by another aircraft during flight, should make an initial Airmiss Report by radio or telephone as soon as possible after the incident and confirm it in writing within 7 days.

519. RT frequencies used by ATC in specific Control Zones, Airways, Terminal Areas and those areas of FIR's which lie outside controlled airspace are to be found in the COM Section, Vol 2, of the UK Air Pilot.

520. (MAP) The symbol Ⓖ signifies a low level Parascending site.

521. On the "Half Million" topographical chart, areas marked in green indicate wooded areas.

522. (SIGNALS AREA) A red letter "L" on the shaft of a white dumbell signifies that Student Pilots may land on grass areas only.

523. A "Rate 2 Turn" is one in which the aircraft turns 360° in two minutes.

524. Wind speed is normally less below 1,000 ft agl than it is above 2,000 ft agl.

525. VOLMET is a continuous VHF broadcast reporting on the actual weather at selected aerodromes.

526. CAP 85 is a summary of the procedures used in RTF.

527. 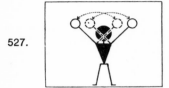 (MARSHALLING SIGNAL) Arms repeatedly crossed above the head, means: ALL CLEAR.

528. Wind shear may occur on all sides of a thunderstorm and the gust front may precede the storm by as much as 20 miles.

529. A kite is an aircraft.

530. Strong turbulence is more likely to be experienced during the passage of a warm front than during the passage of a cold front.

531. Zone Control Units control all take-offs and landings within a particular Control Zone.

532. (MAP) The adjacent symbol signifies the location or a civil aerodrome with a runway or landing strip of 1829 m or over.

533.  An aircraft shall not be flown in simulated instrument flight conditions unless a competent observer occupies a position in the aircraft with a field of vision which makes good the deficiencies of the pilot.

534. Using a dry chemical fire extinguisher inside an aircraft cabin will not affect visibility or breathing.

535. The SALR is approximately the same as the DALR.

536. Induced drag may account for over 80% of the total drag at the low speeds employed during lift-off and landing.

537. (MET) The adjacent symbol indicates: Snow.   ✳

538. There is more mixing between the lower level and higher level air by day than by night.

539. An aeroplane which is banked needs to produce the same lift as weight in order to maintain level flight.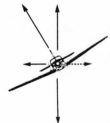

540. If a circuit breaker fuse trips open, the pilot should wait approximately two minutes before re-setting it.

541. The Altitude of the tropopause is higher at the equator than at the poles.

542. (SIGNALS AREA) Two vertical yellow bars on a red square indicates Customs authorities on duty.

543. 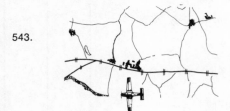 In the UK, an aeroplane flying in sight of the ground and following a line of landmarks (road, railway, canal, coastline, etc.) shall keep such line of landmarks on its right (the Right Hand Traffic Rule).

544. With an altimeter setting of 1013.2 mb, 24,500 ft is not available as a cruising level, it being the plane of division between Quadrantal Rule levels and Semi-Circular Rule levels.

545. (MAP) The numerals shown above the adjacent symbol indicate height above the surface.

546. A pilot does not have to file a Flight Plan before take-off if he intends to land in a different country.

547. The Transverse Mercator Projection is used for the portrayal of large areas of the earth's surface.

548. It is not good airmanship to fly directly underneath any thunderstorm even when you can see through to the other side.

549. At night, bursting projectiles showing white lights or stars directed towards an aircraft mean: LAND AT THIS AERODROME AND PROCEED TO THE PARKING AREA.

550. Balance tabs move in the opposite direction to their elevators, and antibalance tabs move in the same direction as their stabilators.

551. A single arrow on one side of a vector triangle of velocities indicates TRACK and GROUND SPEED.

552.  The London and Scottish Flight Information Regions extend upwards to 30,000 ft above which are their respective Upper Information Regions.

553. If a fuse of higher amperage than that specified for the electrical circuit is fitted, a fire hazard may exist.

554. (MAP) The symbol ⋀⋁⋀⋁⋀⋁ indicates: Cable Joining Obstructions.

555. When flying over mountain areas the minimum safety altitude should be increased by at least 1,000 ft.

556. Whenever the fuel/air mixture has to be leaned, the mixture control should be adjusted to slightly on the lean side of the peak RPM.

557. In anticyclonic conditions, mist, fog and haze are all very rare.

558. (MET) The adjacent symbol indicates: Thunderstorm.

559. **Vsi** is the stalling speed, no power, flaps down and a load factor of I.

560. When carrying out a Special VFR Flight over a built-up area the pilot may be instructed by the ATCU to fly at a height less than 1,500 feet above the highest fixed point within 2,000 feet of the aircraft; but under no circumstances should he fly at a height lower than 1,500 feet or a height which would preclude his gliding clear of built-up areas in the event of engine failure.

SPECIAL VFR

561. In temperate regions Nimbostratus may be found between the surface and about 6,500 ft.

562. On significant Weather Charts, scalloped lines are used to enclose areas of low cloud below 3,000 ft amsl.

563. Hill shading is used to depict relief on the ICAO World Aeronautical Chart Series (1:1,000,000).

564. When an aircraft is flying near to one of the Poles, the angle of dip in the earth's magnetic field will have no discernable effect on the readings of the magnetic compass.

565. 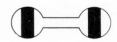 (SIGNALS AREA) A white dumbell with a black strip superimposed in each circular portion signifies that aeroplanes and gliders taking off or landing shall do so on a runway but that movement on the ground is not confined to paved or similar hard surfaces.

566. During a climbing turn the rate of climb will remain constant providing the power remains constant.

567. A dispensation can be granted for student pilots under training to operate aircraft radio without holding a "Flight Radiotelephony Operator's (Restricted) Licence".

568. The carriage of aerosol cans and similar items packaged under pressure will not present a hazard to light aircraft

569. Military Flight Training Areas are situated only above FL 245.

570. A personal flying log book must be kept by all pilots, members of flight crew and persons engaged in flying for the grant or renewal of a licence.

571. When an aircraft is operated contrary to the conditions contained in the C of A, the pilot may be liable to a fine not exceeding £1,000 and/or a term of imprisonment not exceeding 2 years.

United Kingdom
Civil Aviation Authority
CERTIFICATE OF AIRWORTHINESS

572. A flashing white light signal from aerodrome control to an aeroplane in flight means: GIVE WAY TO OTHER AIRCRAFT AND CONTINUE CIRCLING.

573.  (MARSHALLING SIGNAL) Right arm down; left arm repeatedly moved upward and backward, means: Apply right brake.

574. An upslope of 2%, even on a hard surface, will increase the take-off run of a light aircraft by 20%.

575. An upward moving aileron rotates a smaller number of degrees than a downward moving aileron.

576. The spoken word PAN on RTF indicates that the pilot is preparing to enter a TMA and requires navigational instructions.

577. A steady red light signal from aerodrome control to an aeroplane on the ground means: STOP.

578. Lines of Zero magnetic variation are known as "Agonic Lines".

579. The term "wind velocity" refers only to the speed of the wind at the stated altitude.

580. The term 'Hyperventilation' relates to the human body absorbing too much oxygen and as a result dizziness, nausea and blurred vision can occur.

581. The disposable load of an aircraft comprises the usable fuels and oils together with any passengers, luggage and cargo.

582. Except in emergency, it is an offence to drop artices from an aeroplane in flight. (Unless permission has been granted by the CAA.)

583. If the pitot tube becomes blocked by ice, the Altimeter will cease functioning.

584. Aircraft which have demonstrated a satisfactory airworthiness standard, but which are not recognised by the ICAO, are certificated in the Private Category or given a Permit to Fly.

585. Ice crystals or water droplets which reduce visibility at earth's surface to between 1,000 and 2,000 m are called "mist".

586.  Details of aeronautical maps and charts published by the CAA are to be found in the AGA Section of the "UK Air Pilot".

587. Every holder of a medical certificate issued under the articles of the ANO should inform the CAA if he/she suffers any illness involving incapacity to undertake those functions for a period of two weeks.

588. The vents in the tops of fuel tanks are designed to allow any unwanted water vapour to escape without contaminating the fuel.

589. The word "Runway" means an area, whether paved or not provided for the take-off or landing run of aircraft.

590. To obtain a Route Forecast in the form of a Significant Weather Chart, a pilot must always give the meteorological unit at least 4 hours notice.

591. Aeroplanes placed in the Special Category may be used for any purpose specified in the C of A except flying instruction for reward or public transport.

592. Obstructions within 4 nm of an aerodrome reference point are listed in the AGA Section, Vol 2, of the UK Air Pilot.

593. (MET) The letters EMBD on a Significant Weather Chart mean 'Expected moderate base deterioration'.

594. A short exposure to high concentrations of carbon monoxide will not seriously affect a pilot's ability to operate an aircraft.

595. Persons suffering from a head cold should avoid flying in non-pressurized aircraft as during a descent the air pressure either side of the ear drum may not equalize, thus causing pain and possibly damage to the ear drum.

596. The magnetic variation over the UK is easterly.

597. Lenticular clouds indicate the presence of mountain waves.

598. Nimbus clouds are found only at very high altitudes and are composed of ice crystals.

599. A good rule in relation to the consumption of alcohol and flying is to allow at least 8 hours between drinking small amounts of alcohol and flying and where larger amounts have been consumed the period should be much longer.

600. The mean chord line is a straight line connecting the leading and trailing edges of the aerofoil.

Section　2

601. The maximum lift/drag coefficient is achieved by adjusting the airspeed.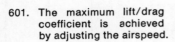

602. (MET) the adjacent symbol indicates "No cloud visible".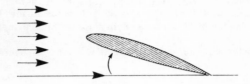

603. Controlled Airspace is subdivided into Control Zones, Airways and Terminal Control Areas.

604. The term **COVAK** means that visibility is 10 km or more, that there is no cloud below 5,000 ft and no cumulonimbus, precipitation, thunderstorms, shallow fog or low drifting snow.

605. Once notice of ETA has been sent to the destination aerodrome, the commander of an aircraft must inform the destination authorities, as quickly as possible, of any estimated delay in arrival of 45 minutes or more.

606. A Certificate of Release to Service is signed by a pilot certifying that he has at all times, while flying the aircraft, complied with the conditions laid down in the C of A.

607.  A double arrow on one side of a vector triangle of velocities is used to indicate WIND VELOCITY.

608. XXX in Morse Code, transmitted by sound (except RTF) or visual signalling, indicates that the pilot requests immediate permission to land.

609. (SIGNALS AREA) Black numerals on a yellow background indicate the direction for take-off or the runway in use.

610. METAR's are Aerodrome Weather Reports.

611.  If the engine temperature is high and the oil pressure is normal, the pilot should lean the mixture and increase the power.

612. The UK is divided into a number of Altimeter Setting Regions (ASR's) in which an area QNH is forecast for every 6 hours.

613. If an aircraft is loaded so that the C of G is towards the aft limit, its stability will be reduced and the effectiveness of the tail surfaces will be reduced.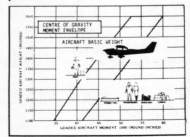

614. A Private Pilot Licence will expire if during a period of 13 months after its initial award or any subsequent renewal the holder fails to complete a minimum of 5 hours flight time as pilot in command.

615. During a descending turn the airspeed will tend to decrease.

616.  In relation to charts 'conformality' is necessary if bearings or angles are to be presented accurately.

617. The atmospheric pressure falls as the warm front approaches.

618. (SIGNALS AREA) A single red letter "L" indicates that only qualified pilots may use the manoeuvring area.

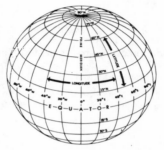

619. The Equator is a great circle but it is not a rhumb line.

620. The law of the UK expressly forbids the carriage of weapons and munitions of war (excluding verey pistols and similar signalling devices).

621. Cold air can contain more water vapour than warm air.

622. (MET) The abbreviation **Cs** means: 'Cirrostratus'.

623. In uncontrolled airspace, at over 3,000 ft amsl, the holder of a PPL may fly with or without passengers provided he can stay 1,000 ft vertically and 1 nm horizontally clear of cloud, and flight visibility is not less than 5nm.

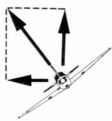

624. A 30° angle of bank increases the stalling speed by 30%.

625. If the pressure and temperature of a given volume of air remains constant, its density will vary in direct proportion to the relative humidity.

626. A steady red light signal from aerodrome control to an aeroplane in flight means: GIVE WAY TO OTHER AIRCRAFT AND CONTINUE CIRCLING.

627. For the first few thousand feet amsl, the atmospheric pressure drops by approximately 1mb per 30 feet.

628. The optimum airspeed for the maximum rate of climb is about 10 knots less than the airspeed for the maximum angle of climb.

629. A flashing white light signal from aerodrome control to an aeroplane on the ground means: MOVE CLEAR OF THE LANDING AREA.

630. ⟍⟍⟍⟍⟍⟍⟍) (MET) The adjacent symbol indicates 'Cirrus'.

631. When determining right-of-way between converging aircraft, an aeroplane towing a glider shall be deemed a single aircraft under the command of the pilot of the towing aeroplane; it will give way to gliders but will have right-of-way over mechanically driven aircraft.

632. On the "Half Million" chart, contour lines mark every 500 ft amsl but layer tinting marks every 1,000 ft.

633. The ground speed of an aircraft is its actual speed over the ground i.e. the TAS corrected for wind velocity.

634. (MET) The abbreviation **FG** signifies "Freezing- Ground Frost".

635. Frise ailerons are designed to increase the drag on the lowered wing.

636. (MET) A **col** is an area of light winds.

637. The number appearing at the top of the C of A is the registration number.

4235

638. **Vfe** is the minimum lift-off speed.

639. If the external static source were to become blocked during flight and if there were no alternate source inside the cabin, the Altimeter and Vertical Speed Indicator readings would remain constant but the Airspeed Indicator would continue to function, though with much less accuracy.

640. Mercator projections are generally used for pilot navigation except in equatorial latitudes.

641. The curvature of the surface of an aerofoil is called its "depth of section".

642. (MET) On a Significant Weather Chart the abbreviation **FRQ** means: "Frequent".

643. A disadvantage of the 1:250.000 Topographical chart is that it does not show the base of Airways.

644. The term "Cloud Ceiling" is the height aal of the lowest part of any cloud, visible from the aerodrome surface, sufficient to obscure more than one half of the sky so visible.

645. In Temperate Regions, high cloud formations (Cirrus, Cirrostratus and Cirrocumulus) are not met with below 25,000 ft amsl.

646. (SIGNALS AREA) A red and yellow striped arrow placed along two adjacent sides of the signals area and pointing in a clockwise direction signifies that a right hand circuit is in force.

647. In Britain, when an aeroplane is flying towards an area of higher pressure the prevailing wind will be from the right and its altimeter will under-read.

648. All engines operate more efficiently and are less likely to overheat when operated at high power with a lean mixture.

649. ↑ In the visual signal code used by survivors, the adjacent symbol means "Am proceeding in this direction".

650. (MAP) ▼ This symbol indicates a Bird Sanctuary.

651. Stratus clouds form in fairly stable atmospheric conditions.

652. Information relating to customs facilities at aerodromes will be found in the **RAC** section of the UK Air Pilot.

653. Aerodrome Control Units control all aircraft:
(a) approaching the aerodrome circuit
(b) within the aerodrome circuit
(c) manoeuvring on the ground.

654. "In Command" flight time refers to times when you are the senior pilot in the aircraft.

655. In the Northern Hemisphere a person standing with his back to the wind will have the low pressure area on his right.

656. The "Air Navigation Order" is one publication through which the CAA promulgates information relating to aviation legislation.

657. (MET) ∽ This symbol indicates "Stratocumulus.

658. Aircraft placed in the Special Category may never be used for Aerial Work.

659. A "Rate 3 Turn" is one in which the aircraft is turning at the rate of 180° per 20 seconds.

660. Distress and Diversion ATC Units of the RAF maintain a continuous watch on 112.5 MHz.

661.

To obtain the Centre of Gravity arm of an aircraft, the total all-up weight must be divided by the Reduction Factor.

662. Radiation Fog is formed by humid air in areas which lose heat by radiation during the night.

663. Light aircraft should allow one mile clearance when flying near any active thunderstorm.

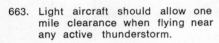

664. In order to comply with the "Quadrantal Rule" a flight level of 180 should be used by aeroplanes crusing on a magnetic track of 180°.

665. (MET) The abbreviation **Ns** means: "No significant weather".

666. Regulations relating to the use of aircraft for agriculture and other aerial work are to be found in the AGA Section, Vol. 2 of the "UK Air Pilot".

667. Detailed information on areas and locations which are restricted for reasons of safety is to be found in the CHARTS Section, Vol. 3 of the "UK Air Pilot".

668. Water vapour weighs less than dry air.

669.

(MARSHALLING SIGNAL) Left arm down; right arm repeatedly moved upward and backward, means "Move Ahead".

670. All air navigation obstructions above 300 ft agl are lighted with a red obstruction light or lights. If at any time these lights should fail, their unserviceability will be promulgated in a Class I NOTAM.

671. Pilots who have obtained the permission of the ATCU concerned to enter a Special Rules Zone or Area, and who have notified the ATCU of their position, level and track, may continue their flight without further mandatory procedures.

672. **N** In the visual signal code used by survivors, the adjacent symbol means: "No".

673. The "One in Sixty" Method of assessing Track Error involves multiplying the Distance Off Track by 60, and dividing the result by the Distance Flown.

674. Runway 08 is in use. The surface wind is 020/20 so the crosswind component is 10 kts.

675. (MET) ▲▲▲ This symbol indicates a "Warm Front".

676. True track plus westerly variation equals magnetic track.

677. The visual code symbol used to indicate "Supplies are required" is shown on the right.

678. With decrease in altitude between 2000' and ground level the winds tend to veer.

679. The Vno, or Normal Operating Speed, is the maximum speed that the aircraft may be flown in turbulent air.

680. Aircraft, Engine and Variable Pitch Propellor Log Books must be kept up-to-date in respect of flight times completed, and destroyed when the aircraft/engine/propeller has been destroyed or withdrawn from use.

681. You are approaching from the south and you report airfield in sight. You are instructed, "G-RN JOIN LEFT BASE RUNWAY TWO TWO" in zero wind conditions your "Base Leg' heading should be 310°.

682. Military Aerodromes notified as available for civil use are listed in AGA 2 in the same manner as civil aerodromes, and such aerodromes may be used in the same manner as civil aerodromes in all respects.

683. Control Zones are established around major airports from ground level to specific altitudes.

684. When a pilot who has filed a Flight Plan diverts, or lands at any aerodrome not specified in his Flight Plan, he is required to inform his destination aerodrome within 30 minutes of his previously planned ETA there.

685. ◕◡ (MET) The adjacent symbol indicates "Variable Winds".

686. In the UK, instrument approach practice is forbidden in VMC, unless the ATCU has been informed and if conducted in simulated flight conditions a competent observer is carried.

687. Air Defence Regulations preclude the publication of detailed information in relation to specific Low Level Routes where military high speed flying takes place.

688. A free balloon is an aircraft.

689. Over the UK the change in the magnetic variation is relatively slow, averaging about one degree annually.

690. Clear ice may form on the exterior surfaces of aircraft descending from sub-zero altitudes through areas of falling rain.

691. A pilot who sees projectiles showing bursting white light or stars, in his vicinity should land at the nearest available aerodrome.

692. (MARSHALLING SIGNAL) Raise arm and hand with fingers extended horizontally across the body, then clenched fist, means: "PULL BACK THROTTLE".

693. When the wet bulb thermometer registers almost the same temperature as the dry bulb thermometer, the relative humidity is high.

694. When flying from an area of lower temperature to an area of higher temperature, the altimeter will under-read.

695. If the owner of an aeroplane in the Special Category chooses to maintain his aircraft in accordance with an approved maintenance schedule, the validity of the C of A may be increased from 2 to 4 years.

696. In relation to "Diurnal effect" by day, surface winds usually tend to veer and increase in strength.

697. "High Density Altitude" is used to describe a situation where the air density is abnormally high at a particular altitude.

698. If you increase your touch-down speed by 5% you will increase your ground roll by 10%.

699. The use of flap increases drag without increasing lift.

700. (MET) The abbreviation **SN** means "Supercooled Nimbus".

701. (MET) The adjacent symbol on a Significant Weather Chart Ψ signifies "Moderate airframe icing".

702.
Aeroplanes placed in the Special Category may not be hired from one person to another if the basic weight is less than 910 kg.

703. At certain Customs Airports, Customs attendance is available only by prior arrangement with the Airport Manager.

704. A small amount of winter morning hoar frost will not significantly affect the performance of a light aircraft.

705. Overprinted aeronautical information relating to Controlled Airspace with lower limits above 3,000 ft amsl is not shown on the (1:250,000) Topographical Air Chart.

706. Pilots who do not hold an Instrument Rating may use Airways without prior clearance provided they do so in Visual Meteorological Conditions.

707. Cumulus clouds form in unstable atmospheric conditions.

708. Magnetic track plus westerly variation equals true track.

709. (MET) On the Significant Weather Chart the abbreviation **CLD** means: "Cloud(s)".

710. Advection Fog is the low cloud often formed by a warm front.

711. 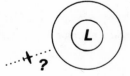 A helicopter is not an aeroplane.

712. The commander of a flying machine on, or in the traffic zone of an aerodrome shall observe such visual signals and marshalling signals as may be displayed and shall obey any instructions which may be given him by means of such signals.

713. Aircraft operating under IFR in Controlled Airspace are not required to conform to the Quadrantral Rule.

714. Bird sanctuaries are not marked on topographical maps unless they extend from the surface to above 4,000 feet agl.

715. Information about the use of airfields—fees and charges; the Public Health, Customs and Immigration regulations; the import and export of goods—are all found in the GEN Section in Vol. 2 of the "UK Air Pilot".

716. When flying towards an area of deteriorating weather, you can expect the aircraft to drift to the right.

717. (MET) The adjacent symbol signifies 'Freezing fog'.

718. (MET) The abbreviation **OVC** means: 'Occasional variable cloud'.

719. If the outside air temperature ,is ten degrees lower than the ISA for that altitude, the altimeter will indicate an altitude 10% lower than the actual altitude.

720. Identification Beacons flash a two-letter morse group every 12 seconds: red for civil aerodromes and white for military aerodromes.

721.  When rising above 'ground effect' the aircraft's nose will tend to pitch upwards.

722. Except for areas of actual precipitation, visibility when flying in cold frontal conditions is better than visibility when flying in warm frontal conditions.

723. Anti-Collision Lights are flashing red lights fitted on aircraft and showing in all directions to attract the attention of other pilots.

724. If the external static source were to become blocked and an alternate static source inside the cabin were to be used, the Airspeed Indicator and the Altimeter would both under-read.

725. (MET) The abbreviation **CAT** on a Significant Weather Chart indicates 'Thunderstorm activity'.

726. The lower end of the white arc shown on an Air Speed Indicator dial indicates the flap down stalling speed.

727. The thickness of a wing at any point is called its "depth of section".

728. If the C. of G. of an aircraft is moved aft it is less likely, if mishandled, to go into a spin, but, once in a spin, it will be more difficult to recover.

729. If the weight of an aircraft is increased the 'load factor' will be increased.

730. The various conditions, to which the C of A is subject, are typed, in detail on the rear page of the certificate.

731. A triple arrow on one side of a vector triangle of velocities is used to indicate WIND VELOCITY.

732. Thunderstorms always move in the same direction as the prevailing wind.

733. Small training aircraft are designed to encounter gust velocities of up to 30 ft per second.

734. All aeroplanes must be weighed and the C of G must be determined before the initial issue of a C of A; and on any other occasion required by the CAA.

735. (MET) This symbol indicates a 'Warm front'.

736. "Cross Country Flight" is defined in the ANO as a flight during the course of which the aircraft is more than 3 nm from the aerodrome of departure.

737. A red pyrotechnic signal from aerodrome control to an aeroplane in flight means: DO NOT LAND, THE AERODROME IS NOT AVAILABLE FOR LANDING.

738.  A "Rate 4 Turn" is one in which the aircraft is turning at the rate of 360° per half minute.

739. Airways are normally 10 nm wide and extend vertically from 3000 agl. to 25,000 amsl.

740. Aero piston engines develop less power when the humidity of the air is low.

741. (MET) The abbreviation **FR** indicates: 'Frontal Rain'.

742. Wind speed and direction at the "surface" are measured at a standard height of 10 metres above the ground.

743. The Dew Point is the temperature at which a body of air is saturated with water vapour.

744. Compass deviation alters with changes in heading and a compass deviation card is fitted in the cockpit.

745. Spinning is not classified as an "Aerobatic Manoeuvre".

746. (MET) The abbreviation **Sc** indicates: 'Stratocumulus'.

747. Small flap settings increase drag more than lift.

748. The "Air Navigation (General) Regulations" is a legal document containing general regulations including those which apply to Public Transport Operations.

749. (MET) GRADU means: Gradually becoming.

750. Permission for Special VFR Flight in Control Zones will always be given to a holder of a PPL without an IMC or Instrument Rating providing air traffic conditions permit.

751. A light aeroplane taking off from an airfield situated 1,000 ft amsl when the temperature is 25°C will need an additional 7.5% in the length of its take-off run.

752. The Lambert Conformal Conic Projection is the most widely used in topographical maps for pilot navigation and for the presentation of meteorological information.

753. When two aircraft are approaching each other approximately head-on; when in the air, each shall alter course to the right; and when on the ground, each shall alter course to the left.

754. When Royal Flights take place within existing controlled airspace, their details will not be promulated by Special RF NOTAM.

755. (MET) The adjacent symbol indicates that the visibility   ⚊⚊
is less than 2 kilometres.

756. When flying beneath a TMA but outside the Control Zone, a pilot should set the altimeter datum to the QNH of any aerodrome within the Control Zone, or below the TMA.

757. The ANO sets out a table of instrumentation and equipment which is considered to be an essential minimum to ensure the required level of safety during different types of flight operation (by day or by night, over land or over sea, for public transport or otherwise, etc.)

758. When by night an intermittent white luminous beam is directed at an aircraft, it means that the aircraft must leave the circuit.

759. In uncontrolled airspace at less than 3,000 ft amsl the holder of a PPL may not carry pasengers unless he can remain clear of cloud and in sight of the surface with a flight visibility of not less than 3 nm.

760.  The adjacent altimeter is indicating: 7,880 feet.

761. You are approaching from the north-west in zero wind conditions and you report airfield in sight. You are instructed "G-RN JOIN RIGHT BASE RUNWAY TWO SIX". Your base leg heading should be 360 degrees.

762. (MET) The adjacent symbol indicates: "Snow Showers".

763. Supercooled water vapour or water droplets in the atmosphere will not cause airframe icing.

764. The term "Night" means the time between half an hour after sunset and half an hour before sunrise at surface level.

765. The wind veers after the passage of a warm front and veers again after the passage of the cold front.

766. When an anti-collision light is fitted it shall only be displayed when an aircraft is on the apron at night with its engine(s) running.

767.  (MARSHALLING SIGNAL) A circular motion of the right hand at head level with the left arm pointing to the starboard engine, means: "OPEN UP STARBOARD ENGINE or TURN TO PORT".

768. The aerofoil section of stabilators and tailplanes are usually symmetrical.

769. When an aircraft has drifted off track, the "Closing Angle" method of calculating the Heading Alteration Required involves dividing the Proportion of Track Flown by the Estimated Closing Angle.

770. (MARSHALLING SIGNAL) Arms extended, palms facing inwards, then swung from the extended position inwards, means: "FLAPS DOWN".

771. (SIGNALS AREA) A checkered flag or board containing 12 equal squares, coloured red and yellow alternately, signifies "Racing in progress no unauthorised aircraft may take off or land".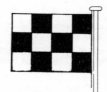

772. In "AIRMET" forecasts the levels of clouds are given in altitudes.

773. A decrease in air density will produce less lift and more drag.

774. A Lower Airspace Radar Advisory Service is available to all aircraft in uncontrolled airspace within 30 nm of each participating airfield when flying between 3,000 ft amsl and FL 80.

775. The commander of an aircraft, on request by a competent authority, must produce all aircraft documentation "within a reasonable time" and all flight crew licences (at a UK police station) within 5 days.

776. 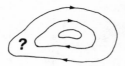 Above 2,000 ft agl the wind will follow the direction of the isobars quite closely.

777. The Minimum Safety Altitude is generally accepted as being 1,000 ft above the highest ground within 2 nm either side of the aircraft's track and around the point of departure and the destination.

778. A 60° angle of bank increases the stalling speed by 40%.

779. The AGA section of the UK Air Pilot lists all TORA, TODA ED and LDA distances in metres.

780. In the code designation of Restricted, Danger and Prohibited Areas, the numbers after the oblique are the Flight Level below which the particular area may not be overflown.

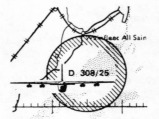

781. The forcast QNH is the average expected reading for the area during the period of validity.

782. A meridian is half of a great circle but it is not a rhumb line.

783. An aircraft shall not fly closer. than 500 feet to any person vessel, vehicle or structure.

784. (SIGNALS AREA) A double white cross and/or two red balls suspended from a mast one above the other signifies that glider flying is in progress at the airfield.

785. The category in which an aircraft may be operated is shown on its Certificate of Registration.

786. The "Track True" of an aircraft is its intended path over the ground measured from true north and allowing for variation and deviation.

787. (MARSHALLING SIGNAL) Right or left arm down, other arm moved across the body and extended to indicate position of the other marshaller, means "PROCEED UNDER THE GUIDANCE OF ANOTHER MARSHALLER".

788. A flashing green light signal from aerodrome control to an aeroplane in flight means: RETURN TO/OR REMAIN IN THE CIRCUIT AND AWAIT A SIGNAL FOR LANDING CLEARANCE.

789. By night, surface winds usually tend to veer and increase in strength.

790. If the oil pressure gauge reads zero or very low and the oil temperature and/or cylinder head temperature remains normal, the pilot should give consideration to the fact that the oil pressure gauge could most likely have become unserviceable.

791. A pilot who wishes to operate within a Special Rule Zone is exempt the need to obtain an ATC clearance if he is operating in accordance with the Visual Flight Rules.

792. If the temperature is 15°C at sea level, it will be approximately 10°C at 5,000 ft and 5°C at 10,000 ft.

793. (MET) The adjacent symbol indicates: "Cumulonimbus".   ⌣

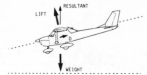

794. The glide ratio of an aircraft is equal to its lift/drag ratio.

795. The British Isles are covered by 3 F.I.R's (London, Preston and Scottish) each of which can be contacted on a specific frequency.

796. **Va** is the design manoeuvring speed.

797. In a spin, the airspeed drops sharply and the pilot's first reaction should be to apply more power.

798. Runway 08 is in use. The surface wind is 060/20 so the crosswind component is 7kts.

799. In straight and level flight, in calm air, an aeroplane is subjected to a load factor of zero.

800. For the purpose of renewal of the privileges of a PPL, the equivalent of 5 hours flight time in command may be obtained by completing 10 hours dual with a qualified instructor.

801. Prior to the issue of a Private Pilot's Licence, the CAA require evidence of the following training:-
    (a) Not less than 20 hours dual including 4 hours Introduction to Instrument Flying and 4 hours Dual Navigation.
    (b) Not less than 10 hours solo including 4 hours Navigation.
    (c) A minimum total of 40 hours.

802. Before making a flight from any aerodrome with an ATCU the pilot must file a Flight Plan.

803. There are slight variations in scale between the Standard Parallels of a Lambert's chart, but all angles and bearings are accurately portrayed.

804. A licenced pilot who operates and/or owns an aircraft for which he is entitled to carry out maintenance functions, can find the list of maintenance tasks he is legally empowered to carry out, in the "Air Navigation Order".

805. If the Relative Humidity is 30%, this means that the air contains only 30% of the water vapour which it could absorb at that temperature.

806. Two or more white crosses displayed on a runway or taxiway with their arms at 45° to the centre line indicate the position of the Holding Point.

807. The term "Rolling Resistance" is used to denote the coefficient of friction when an aeroplane is rolling on different types of surface.

808. Below 2,000 ft agl the wind will follow the direction of the isobars more closely over the land than over the sea.

809. In a low pressure area, the air is ascending.

810. (MET) The adjacent symbol indicates: "Heavy Continuous Rain".

811. A flashing green light signal from aerodrome control to an aeroplane on the ground means: "CLEAR TO TAXI, or, YOU MAY MOVE ON THE MANOEUVRING AREA".

812. Information relating to its period of validity is to be found in the margin of an aeronautical chart.

813. An aircraft shall not fly within 1,000 ft of any assembly of 3,000 persons or more witnessing or participating in an organised event unless the pilot has the written permission of the CAA and the organisers.

814. (MARSHALLING SIGNAL) Right arm raised at the elbow, with the arm facing forward, means: "STOP".

815. A glider when flying at night shall display a steady red light showing in all directions and no other lights.

816. The wings of light aircraft are designed to prevent the outer sections stalling before the inner sections.

817. Aircraft classified as Special Category Class 1 will not be eligible to operate in the public transport role.

818. The Meteorological Office will issue a Gale Warning when winds in excess of 30 knots are expected near the surface.

819. The aspect ratio of a wing is calculated by dividing the dihedral angle by the length of the wing.

820. A parallel of latitude is a great circle but it is not a rhumb line

821. Runway 36 is in use. The surface wind is 030/24 so the crosswind component is 12 kts.

822. Most of the Range and Endurance tables/graphs shown in aircraft manuals are calculated on the basis that during flight the mixture control will be used.

823. True track minus easterly variation equals magnetic track.

824. Terminal Control Areas (TMAs) are established at any junction between airways.

825. The Basic Weight of an aircraft is the "dry" weight and does not include any items of equipment.

826. An Area Forecast is normally issued every 6 hours commencing from 0001 hours.

827. On a Significant Weather Chart scalloped lines are used to show areas of similar weather conditions.

828. All aircraft intending to enter Controlled Airspace other than for the purpose of take-off or landing must file a flight plan to the appropriate ATCU prior to the intended flight and make Position Reports in accordance with flight clearance instructions.

829. An aircraft voltage regulator controls the rate at which the electrical services take current from the battery.

830. Private pilots who hold an "Instrument Rating" are permitted to enter Control Zones.

831. The presence of stratus clouds indicates smooth flying conditions.

832. The use of "Carburettor Heat" reduces fuel consumption.

833. All aircraft manuals list the distances required for take-off and landing in metres.

834. In relation to Danger Areas, "day" means from 0800 to 1800 local time.

835. Any of the instructional flying necessary for a person to obtain a licence must be conducted from a licensed aerodrome.

836. The AGA Section of the UK Air Pilot gives details of major airports and licensed aerodromes as well as information on unlicensed aerodromes and military aerodromes available for civil use, but not private landing strips.

837. The adjacent altimeter indicates: 9,800 feet.

838. In uncontrolled airspace at less than 3,000 ft amsl the holder of a PPL (without an IMC Rating) may not fly, even without passengers, unless he can remain clear of cloud and in sight of the surface with a flight visibility of not less than 3 nm.

839. Survivors of a forced landing, on seeing a SAR plane's green pyrotechnic signal, should (if available) immediately fire an answering red pyrotechnic.

840. The angle of incidence of the tailplane is usually slightly less than that of the wings.

841. When the holder of a PPL has not completed the necessary flight time within the statutory 13 months, he should undergo a flight test before his privileges can be renewed.

842. (MET) The abbreviation **LYR** on a Significant Weather Chart means: "Layer".

843. When rising above ground effect, the aircraft's acceleration will suddenly decrease.

844. By day, a Sea Breeze often blows from the sea to the land.

845. 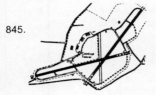 The airfield you are approaching has three runways: 04/22, 08/26 & 18/36. You are instructed "G-RN JOIN LEFT BASE RUNWAY 04". Your correct base leg heading will be approximately 310 degrees.

846. The temperature and the dew point will increase after the passage of a warm front.

847. All limiting load factors are positive.

848. Aerodrome Traffic Zones have a horizontal radius which varies dependent upon the length of the main runway.

849. Frontal Fog is sometimes formed by the passage of a Warm Front or a Warm Occlusion with cloud slowly lowering to the surface.

850. During the hours of operation of an aerodrome Air Traffic Unit a pilot need not comply with the directions issued by Air Traffic Control unless this requirement is specified in his air traffic clearance.

851. Restricted Areas can only be entered with the permission of the controlling ATC unit.

852. High temperatures and high humidity reduce both lift and thrust during take-off.

853. The ICAO World Aeronautical Chart Series (1:1,000,000) have too large a scale to be interpreted easily during flight.

854.  (MARSHALLING SIGNAL) Either arm and hand placed level with the chest, then moved laterally with the palm downwards, means: SLOW DOWN.

855. When Light Aircraft are used in the Public Transport Category the servicing periods are calculated in both calendar months and flying hours: a typical figure being 1,000 hrs or six months (whichever is the sooner).

856. When the Dry Bulb temperature is close to the Wet Bulb temperature and light winds prevail, conditions will be conducive to the formation of fog.

857. Any Aircraft, may, without ATC clearance, fly at right angles across the base of an en-route section of an Airway where the lower limit is defined as a Flight Level.

858. (SIGNALS AREA) A red square panel with a yellow stripe along each diagonal signifies that the state of the manoeuvring area is poor and that pilots must exercise special care when landing.

859. The "Manoeuvring Area" comprises the areas provided for take-off and landing and other movement on the surface excluding the apron and maintenance areas.

860. When flying towards an area of deteriorating weather conditions, your altimeter will begin to under-read.

861. The standard Transition Altitude for civil aerodromes outside Controlled Airspace is 2,000 ft QNH.

862. An aeroplane placed in the Transport Category may be operated for Public Transport, or private use.

863. The RAC Section (Vol 1) of the UK Air Pilot contains the "Rules of the Air and Air Traffic Control Regulations".

864. Calibrated Airspeed is the American equivalent of Rectified Airspeed.

865. The aircraft Flight Manual forms part of the Certificate of Airworthiness.

Flight/Owner's Manual.
Pilot's Operating Handbook.

866. The term "drift" means the angular difference between the track required and the actual track made good.

867. (MET) The abbreviation SCT means: "Severe cloud turbulence".

868. (MET) The adjacent symbol signifies an "Occlusion". ▲▲▲

869. Low density air produces less lift and less drag.

870. A climb from sea level to 3,000 ft will result in a larger drop in pressure in warm air than in cold air.

871. Magneto's are used to provide an electric current which creates: electrical power for the starter motor.

872. Airframe icing may extend over the engine's external air intake and result in partial or even complete loss of power

873. (MET) The abbreviation TS means: "Turbulence severe".

874. An overtaking aeroplane shall: when in the air, alter course sufficiently to the right to pass well clear of the aeroplane being overtaken; when on the ground, alter course sufficiently to the left to pass well clear of the aeroplane being overtaken.

875. In the absence of Identification Beacons, Aerodrome Beacons flash a white/green or white light.

876. A level turn at a bank angle of 30° increases the load factor to more than 1.5.

877. When approaching large cumulus clouds, the safest course is to fly over the top of them.

878. If, by mistake, a pilot sets the QNH on his altimeter 10 millibars higher than its actual value, he will be under the impression that his altitude is is about 300 ft higher than it really is.

879. If the C of G of an aircraft is moved forward, the longitudinal stability is increased.

880.

(MARSHALLING SIGNAL) Arms down, the palms facing outwards, then swung outwards, means: "SLOW DOWN".

881. In the traffic zone of an aerodrome where no ATCU is, for the time being notified on watch, aircraft may neither take-off nor land except in emergency.

882. The lower the atmospheric pressure at ground level, the greater the lift developed by the wings during take-off.

883. The '½ million' charts used in pilot navigation use True North as a reference datum.

884. (MET) The abbreviation AC means: "Altocumulus".

885. A red pyrotechnic light or flare signal from an aircraft to aerodrome control means: "IMMEDIATE ASSISTANCE IS REQUIRED".

886. The International Standard Atmosphere is based upon the mean meteorological conditions existing at Greenwich.

887. "Notified" means, promulgated in a document entitled "Notam UK" or "UK Air Pilot".

888. When the Carburettor Hot Air system is in use the induction air filter is normally by-passed.

889. **Vno** is the never exceed speed.

890. An airship is not an aircraft.

891. (MET) The term **Tempo** means: "Gradually Increasing".

892. Both lift and all forms of drag (except vortex drag) increase in proportion to the square of the airspeed.

893.  The carriage of radio is mandatory for private or training flights only if such flights are made at or about FL245 or near certain airfields surrounded by Controlled Airspace or Special Rules Airspace.

894. The adjacent symbol shown on Significant Weather Charts means: "Expected movement and speed of the associated front or centre of pressure". **25**

895. (SIGNALS AREA) A white "T" indicates that the direction of the wind is parallel with the shaft of the "T" and towards its crossarm.

896. The Attitude Indicator (Artificial Horizon) is operated by an engine driven pump or an external venturi system.

897. Information relating to Nationality and Registration Marks of aircraft, Civil Aviation Legislation and Air Navigation Regulations are all to be foud in the RAC Section. Vol. 1 of the "UK Air Pilot".

898. When the altimeter datum is set at local QNH, the reading will show the altitude agl.

899. Pilots holding the IMC Rating may be granted Special VFR Flight permission in Control Zones and some Special Rules Airspace provided visibility is not less than 1.5 nm and weather and air traffic conditions are suitable.

SPECIAL VFR

900. High Intensity Radio Transmission Areas are shown on topographical charts and their vertical limits vary.

901. AGA Vol. 2 of the UK Air Pilot lists all customs airports which may be used for entry to, or departure from, the UK.

902. (SIGNALS AREA) A red square panel with a yellow diagonal strip signifies that the airfield is unsafe for the movement of aircraft and that landing on the airfield is prohibited.

903. Buys Ballot's Law:- If you stand with your back to the wind in the Northern Hemisphere, the low pressure will be to your left.

904. Initial test, servicing and acceptance flight tests excepted, an aeroplane without a valid, current C of A may not be flown.

905. The basic weight of an aircraft is the weight of the aircraft, the equipment necessary for its certification, its unusable fuel together with either its unusable oil or, if specified, its usable oil.

906. (MET) **OCNL** means: "Overcast cumulonimbus layered".

907. **An aeroplane will glide further when take-off flap is selected.**

908. A level turn at a bank angle of 60° increases the load factor by 60%.

909. The colour of AVGAS LL is blue.

910. An aeroplane placed in the Private Category must have its C of A renewed every 2 years.

911. The pilot must ensure that the aircraft's Certificate of Airworthiness is valid prior to take-off.

912. Where sufficient hygroscopic nuclei are present, cloud formation can take place when the relative humidity is as low as 85%.

913. Indicated Airspeed, when corrected for instrument and position error, is known as True Airspeed.

914. The average Lapse Rate is about 5°C per 1,000 ft increase in altitude.

915. "Track error" is the actual path of the aircraft over the surface as distinct from the intended track to be flown.

916. By mishandling, a pilot can subject a modern light aircraft to unsafe load factors without 'blacking' himself out.

917. The filing of a Flight Plan is required before flights conducted under IFR in Controlled Airspace.

918. (MAP) The adjacent aeronautical chart symbol   ▬ • ▬   means: "International Boundary".

919. ⌒ (MET) On a Significant Weather Chart the adjacent symbol means: "Mountain Waves".

920. When a quantity of air is compressed into a smaller volume it becomes warmer.

921. When the altimeter datum is set at QFE, the reading will show the height above Mean Sea Level.

922. The operation of the aircraft outside the conditions in the C of A will automatically render the C of A invalid.

923. If the wind is from dead ahead on take-off it will normally be from several degrees to the left at 2,000 ft agl.

924. The various scales of mandatory emergency equipment for Private Flights are to be found in a Schedule of the ANO.

925. Permanent (Scheduled) Danger Areas are operational for H24 and shown on the "Chart of UK Airspace Restrictions" with an unbroken red outline.

926. The sublimation of hoar frost on the exterior surfaces of aircraft can occur when an aircraft, after flying in sub-zero temperatures, descends through warmer air with a higher relative humidity.

927. If the holder of a PPL has not flown as pilot in command for 26 months since the expiry of his privileges, he will have to complete one hour dual, one solo flight and pass a flight test including pre-flight-planning.

928. An aeroplane must not fly over any congested area of a city, town or settlement below such a height as would enable it to be flown clear of the area in the event of engine failure, or below 3,000 ft amsl, whichever is the higher.

929. Carburettor icing takes place only when the relative humidity is high and the outside air temperature is below freezing.

930. (SIGNALS AREA) A white disc displayed above the crossarm of a white "T" and/or a black ball suspended from a mast signifies that the directions of landing and take-off do not necessarily coincide.

931. Short grass on firm ground will increase the take-off run of a light aircraft by about 7%.

932. The CAP 53 is the Controlled Approach Procedure as approved by the CAA in 1953.

933. Recovery from a spin depends first on controlling the pitch.

934. (MARSHALLING SIGNAL) Raise arm, with fist clenched, horizontally across front of the body, then extend fingers, means: RELEASE BRAKES.

935. RAF and RN "Master Aerodromes" are operational H24 each day and their Identification Beacons will operate continuously throughout the hours of darkness.

936. An accident shall be notified by the commander of an aircraft if, between the time when any person boards the aircraft with the intention of flight and such time as all persons have left the aircraft, any person suffers death or serious injury while in, on or in contact with the aircraft or anything attached to it.

937. The regulations pertaining to Certificates of Test and Experience are contained in the "Rules of the Air and Air Traffic Regulations".

938. The "Order" means the order in which aircraft currently in the aerodrome circuit will have permission to land.

939. The difference between True and Magnetic headings is known as "Magnetic Deviation".

940. The adjacent aeronautical chart symbol means: 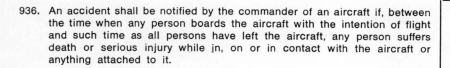 "FIR Boundary".

941. Glider sites may have winching operations extending to 1,500 ft above the surface.

942. The Certificate of Approval of Aircraft Radio Installation relates to the radio equipment and signifies that the radio installation complies with BCAR's.

943. Runway 36 is in use. The surface wind is 350/30. Therefore the cross-wind component is 10 kts.

944. Vne is the maximum safe speed with flaps extended.

945. Certain military aerodromes available for civil use have a Transition Altitude above 3,000 ft and pilots using these aerodromes should use the aerodrome QFE.

946. (MET) The adjacent symbol means: 'Fog increasing'.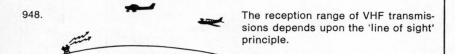

947. It is not necessary to file a Flight Plan for a Special VFR Flight but prior clearance from the appropriate ATC Unit must be obtained.

948.  The reception range of VHF transmissions depends upon the 'line of sight' principle.

949. (MET) The adjacent symbol signifies: 'Rising Stratus'.

950. A Lambert Conformal Projection produces great circles which are concave towards the nearer pole.

951. The Air Traffic Advisory Service is available to aircraft flying under IFR in Advisory Service Areas or along Advisory Routes.

952. The "Mass Balance" method of reducing the possibility of control surface flutter involves the addition of weights placed in a manner which moves the C of G of the control surface aft of the hinge line.

953. Stratus clouds may cover a very large area and have a vertical extent of many thousands of feet.

954. The density of the air is inversely proportional to the atmospheric pressure.

955. A special VFR Flight clearance cannot be issued to any Private Pilot who does not hold an Instrument Rating.

956. On Lambert Conformal Conic Charts all Meridians converge towards the nearest pole.

957. A white cross displayed at the end of a runway indicates that landing is dangerous and that the airfield is used only for storage.

958. (MET) The abbreviation **INTER** means: The meteorological conditions between the higher altitude and lower altitude.

959. Magnetic track plus easterly variation equals true track.

960. Wings with a high aspect ratio have a higher stalling angle than wings with a low aspect ratio.

961.  At aerodromes not having an ATCU on watch and at which take-offs and landings are not confined to a runway, a flying machine or glider when landing shall leave clear on its right any aircraft which has already landed, is already landing or is about to take-off.

962. An aircraft signalling by white flashes, switching on and off its landing and/or navigation lights means: I AM COMPELLED TO LAND.

963. A steady green light signal from aerodrome control to an aeroplane in flight means: RETURN TO THE AERODROME AND AWAIT PERMISSION TO LAND.

964. By night, a Land Breeze often blows from the land to the sea.

965. All Private Aircraft required by the owner solely for private flying will be placed in the Special Category.

966. Although turbulence may surround large cumulus clouds, the air just beneath such clouds is turbulence-free.

967. If Meteorological Warnings, e.g., Gales, Frost, Snow etc. are required by Aerodrome Operators, application must be made to the Meteorological Office.

968. A representative of the CAA or any other authorised person may enter and inspect an aircraft to determine whether the intended flight will be in contravention of the ANO.

969. If the Alternator/Generator fails during flight, the Magnetos will cease to function when the battery becomes discharged.

970. As a general rule, the Centre of Pressure moves rearwards when the angle of attack is increased.

971. In a ridge of high pressure, the air is ascending, expanding and cooling.

972. (MET) The adjacent symbol is used to indicate: "No Significant Weather".

973. The minimum age for the issue of a Private Pilot's Licence or the award of a Student Pilot's Privileges is 17 years.

974. The ICAO World Aeronautical Chart Series (1:1,000,000) is based on the Lambert's Conformal Conic Projection.

975. On a Mercator chart, areas become increasingly distorted away from the Equator.

976.  An altimeter's accuracy is acceptable if it indicates within one hundred feet, plus or minus, of the true altitude; if it indicates a greater error, in either direction, it must be reported as unserviceable.

977. If the carburettor ices up repeatedly, only the lowest throttle setting should be used.

978. Inside controlled airspace, certain Advisory Areas and Routes have been established and a specific ATC service is available on request.

979. Rules of the Air and Air Traffic Services are to be found in the RAC Section in Volume 1 of the "UK Air Pilot".

980. (MAP) The adjacent symbol indicates: "Gliding is an infrequent activity at this location".

981. The correction of any errors discovered in already published Aeronautical Charts takes place every three months in the amendment sheets of the UK Air Pilot.

982. The aspect ratio of a wing may be determined by multiplying the tangent of the dihedral angle by the square of the span.

983. When flying away from an area of deteriorating weather, you expect your altimeter to begin to under-read.

984. The temperature and the dew point falls when the cold front passes, but the pressure rises rapidly after it has passed.

985. Carburettor Icing can occur when flying in or out of cloud or with outside air temperatures above 0°C.

986. A headwind component of 20% of the lift-off speed reduces the ground roll by 15%.

987. Warm, humid air at ground level permits shorter take-off runs.

988. By telephoning the "AIRMET" Service at West Drayton or Broughton a pilot can listen to pre-recorded forecasts for different areas of the UK.

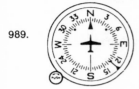

989. An operating Heading Indicator in a stationary aircraft would have to be realigned periodically because of the earth's rotation.

990. A flashing red light signal from aerodrome control to an aeroplane in flight means: DO NOT LAND. WAIT FOR PERMISSION.

991. On Aeronautical Charts, special access lane entry/exit points are depicted by the adjacent symbol.

992. (MET) On significant Weather Charts the adjacent symbol indicates: "A Cold Occlusion".

993. Pilots of aeroplanes who wish to cross an Airway by day in VMC must obtain prior clearance before take-off or, when airborne, at least ten minutes before the intended point of crossing. This clearance is normally obtained by RTF via the FIS.

994. The ANO requires the pilot to take into account the likely weather en route, the weather at destination and need for an alternative route and/or destination.

995. When "QNH" is set on an altimeter the term "Height" should be used to describe the aircraft's vertical distance from the surface.

996. In uncontrolled airspace, not Notified to the contrary, whilst under IFR, at above 3,000 ft amsl, the aircraft must not be flown at a height of less than 1,000 ft above any obstruction within 5nm of its position unless authorised to do so by a competent authority.

997. A free balloon while flying at night shall display a steady white light, showing in all directions, suspended between 5 and 10 metres beneath its lowest part.

998. An aircraft while landing or on final approach to land shall have the right-of-way over other aircraft in flight but not such aircraft as are aready on the ground or water.

999. In so far as Customs requirements are concerned, flights to the Channel Islands and the Isle of Man are considered as flights abroad.

1000.

(MARSHALLING SIGNAL) Arms placed above the head in a vertical position, means: STOP.

1001. Any line which crosses all meridians at a constant angle is called a "rhumb line".

1002. Moist air has a higher density than dry air.

1003. Control surfaces are sometimes fitted with horn balances in order to reduce the force required by the pilot.

1004. A flashing red light signal from aerodrome control to an aeroplane on the ground means : MOVE CLEAR OF THE LANDING AREA IMMEDIATELY.

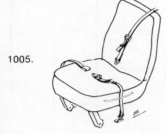

1005. When using non-inertial harnesses the lower torso straps should be tightened before the upper torso straps.

1006. When flying in high pressure areas (Anticyclonic conditions), one normally experiences strong winds and poor visibility.

1007. The term "Controlled VFR Flight" is applicable to all aircraft obtaining permission to enter a SRZ.

1008. A "Rate 1 Turn" is one in which the aircraft turns 180˚ in one minute.

1009. When an aircraft is operated contrary to the conditions contained in the C of A, the manufacturer's warranty, the aircraft's insurance and the pilot's insurance will normally become void.

1010. When by day, in the UK, black or white smoke projectiles are directed towards an aircraft every ten seconds, it means that the aircraft is in a MATZ and immediate RTF contact is required.

1011. The pre-recorded forecasts of the 'AIRMET' Service are up-dated 4 times daily.

1012. **The variable load of an aircraft is the weight of the crew and their** equipment.

1013. Katabatic winds are formed by warm air overflowing sun warmed valleys and displacing cooler air on higher ground.

1014. The angle of attack is the angle between the relative airflow and the lower surface of the wing.

1015. Roll clouds form on the windward sides of mountains.

1016. Authorisation from the appropriate ATCU to make a Special VFR Flight in a Control Zone absolves the pilot from complying with IFR.

1017. 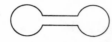 (SIGNALS AREA) A white dumbell signifies that movements of aircraft and gliders on the ground shall be confined to paved, metalled or similar surfaces.

1018. Radio licences are not required for aircraft radio installations.

1019. When a particular aircraft's weight and balance calculations are known for a basic load, a weight and balance calculation for other loads need not be made.

1020. Wings with a high aspect ratio will cause less vortex drag than wings with a low aspect ratio.

1021. When two or more aircraft are approaching a place to land, the aeroplane to the right shall have right-of-way unless the ATCU has given priority **to another aircraft.**

1022. Requests for Special VFR clearance to enter or transit a Control Zone may be made while airborne, giving point of entry and ETA at least 10 minutes before reaching the Zone boundary.

1023. Vso is the stalling speed, no power, flaps down and a load factor of 1.

1024. On pilot navigation maps, all lines joining places of equal magnetic variation are called "Agonic Lines".

1025. (MARSHALLING SIGNAL) Arms repeatedly moved upward and backward, beckoning onward means: RELEASE BRAKES.

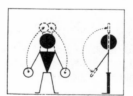

1026. The expansion space at the top of fuel tanks is to allow for an increase in temperature after filling.

1027. **1536** Air Navigation Obstructions reaching 300 ft or more agl, when known to the CAA, are listed in the RAC **(1300)** Section, Vol. 1, of the UK Air Pilot.

1028. True Airspeed is the actual speed of the aircraft over the surface.

1029. (MET) The adjacent symbol indicates a cyclone.   ~]

1030. (MET) CU means : cumulus.

1031. Vra is a specifically recommended airspeed for flight in turbulent air.

1032. A steady green light signal from aerodrome control to an aeroplane on the ground means: YOU MAY MOVE ON THE MANOEUVRING AREA.

1033. Information about the aeronautical telecommunications system is to be found in the COM Section in Vol. 2 of the "UK Air Pilot".

1034. SIGMET is broadcast through the FIR service whenever any kind of hazardous weather conditions are reported.

1035. On a Mercator chart all rhumb lines will appear as straight lines.

1036. The density of the air is directionally proportional to its Absolute Temperature.

1037. Water droplets or ice crystals which reduce visibility at the Earth's surface to less than 1,000 m are called "fog".

1038. The Agonic Line of Zero Variation passes through the Greenwich Observatory in London.

1039. When flying away from an area of deteriorating weather conditions, you can expect a right drift.

1040. The two FIRs covering the UK and the surrounding waters are the London FIR and the Edinburgh FIR.

1041. (MARSHALLING SIGNAL) Arms are placed down with the palm towards the ground, then moved up and down several times, means : SLOW DOWN.

1042. Outside Controlled Airspace in ADR's an advisory service offers a continuous separation service between all participating aircraft but not from other aircraft.

1043. When the Standard Setting is used, the altimeter reading will indicate a Flight Level.

1044. On routine flights in the UK, (with no aerobatics) a drift error on the Heading Indicator of 10° per 15 minutes may be expected.

1045. For flights within notified SRZ, special VFR Clearance from the appropriate ATCU is required, whatever the meteorological conditions.

1046. Experimental Aircraft of all kinds are only certificated in the Special Category Class II.

1047. After an outward bound aircraft has been cleared from a Customs Airport it must not land again in the UK without contacting either the Customs or the Police as quickly as possible.

1048. Contours, spot height and layer tinting form the basis of depicting relief on UK topographical charts. Hill shading may also be shown on the latest issues of the 1:250,000 series.

1049. A aeroplane placed in the Private Category may also be used for Aerial Work.

1050. As a general rule, the higher pressure area will be to the right when you are taking-off into the wind in the Northern Hemisphere.

1051. Captive balloons and kites, flying above 60 metres at night, shall display groups of red and white lights: a triangle of flashing lights on the ground around the mooring point and pairs of white-over-red lights at specified distances up the cable.

1052. (MAP) The adjacent symbol indicates a VOR.

1053. The take-off run will normally be more affected by the type of surface, i.e., paved runway or grass, than by the existing wind velocity.

1054.    As a general rule, the Centre of Pressure moves forward when the angle of attack decreases.

1055. A white letter "L" on the manoeuvring area indicates that part which may be used for landing only.

1056. In Temperate Regions, medium cloud (Altostratus and Altocumulus) is found chiefly between 6,500 ft and 23,000 ft.

1057. (MET) The abreviation CAST signifies castellanus.

1058. When ground features are partially obscured by a layer of mist or haze, it is advisable for the pilot to fly lower in order to discern the ground ahead as well as what is directly beneath the aircraft.

Anti-Clockwise Change
BACKING

1059. When descending from 2,000 ft agl to ground level, you may expect the direction of the wind to back.

1060. Both the Airspeed Indicator and the Vertical Speed Indicator are activated by a combination of pressures from the pitot and static lines.

1061. A Statute Mile equals 5820 feet and a Nautical Mile equals 6280 feet.

1062.    The ground beneath the approach path is lower than the landing runway. In moderate to strong winds a pilot should expect to find a significant downdraught in the final stages of the approach.

1063. The hours of watch for AIS Centres are increased during the summer months.

1064. Light aircraft engines develop maximum power in cold, high density air.

1065. A pilot is not required to file a Flight Plan when he wishes to use an Air Traffic Advisory Service.

1066. A tailwind component of 20% of lift-off speed increases the ground roll by 25%.

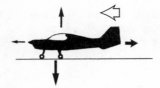

1067. (MET) The letters FU on a Significant Weather Chart mean fog uncertain.

1068. Meteorological units require a minimum of 2 hours notice before issuing a Significant Weather Chart for a Route Forecast of less than 500 nm and 4 hours notice for a Route Forecast covering more than 500 nm.

1069. A Class 3 medical certificate is required to qualify for the privileges of a Student Pilot or Private Pilot's Licence and this will last for 3 calendar years when the holder is under 40 but for only 1 calendar year when the holder is over 40.

1070. (MET) The adjacent symbol signifies: Hail. ▲

1071. The Volmet Frequency is the same for the London and the Scottish FIR's.

1072. In the UK Temporary Danger Areas may be defined and notified around disaster areas.

1073. Notified Danger Areas are marked with a pecked red outline in the "Chart of UK Airspace Restrictions". Their sporadic operational activity is notified by NOTAM.

1074. Cumulus clouds never occur in areas covered by stratus clouds.

1075. When a quantity of air expands into a larger volume it becomes cooler.

1076. Some aerodromes which have Control Zones also have Special Rules Zones.

1077. CAP 85 contains a summary of Aviation Law relating to the Registration and Licencing of Aircraft.

1078. An aircraft in grave and imminent danger requesting immediate assistance should make any one, or a combination of more than one, of the following signals:—
 (a) the signal MAY DAY in Morse Code by visual signalling.
 (b) a succession of single red pyrotechnic lights fired at short intervals.
 (c) A parachute flare showing a single white light.

1079. An aircraft shall not fly closer than 500 ft to any person, vessel, vehicle or structure.

1080. The density of a given volume of air varies in inverse proportion to its pressure.

1081. Cumulonimbus clouds do not occur below 10,000 feet.

1082. (MET) The symbol 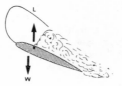 on a Significant Weather Chart indicates: Severe Turbulence.

1083. The point on the wing surface at which smooth airflow breaks down into turbulent flow is called the 'Separation Point'.

1084. (MET) The symbol ≡ indicates: Mist.

1085.  Fire extinguishers containing carbon tetrachloride must never be used in aircraft cabins.

1086. To achieve a rate 1 turn at any airspeed, a 15° angle of bank must be selected.

1087. (MET) The letter RA used in a TAF indicates: Rain.

1088. Water droplets can often remain in liquid state at −10° C.

1089. When conforming to the Quadrantal Rule a pilot flying on a magnetic track of 100° may cruise at any altitude between 3,500 ft and 4,500 ft amsl.

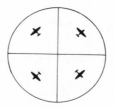

1090. Pilots actively engaged in flying will suffer no ill effects by becoming blood donors.

1091. '(MET) The symbol ⌓ denotes: Thunderstorm with anvil.

1092. Holders of Private Pilot Licences who have not flown as pilot in command for 4 years are required to surrender their licences to the CAA.

1093.  A white cross and a single white bar displayed at the end of a runway at a disused airfield indicates that runway is fit for emergency use though not safeguarded and perhaps temporarily obstructed.

1094. AGA 2, Vol 2, of the UK Air Pilot covers the facilities, runways, local flying instructions, etc., of listed aerodromes.

1095. During an occlusion any build-up of cumulonimbus is easily discernable and the resulting isolated thunderstorms can easily be detected.

1096. The term "Ground Visibility" means the horizontal visibility at ground level.

1097. The specifications of the International Standard Atmosphere at mean sea level include an atmospheric pressure of 1012.5 mb and a temperature of 15° C.

1098. The Topographical Air Charts of the United Kingdom (1:250,000) are based on the Lambert's Conformal Conic Projection.

1099. On a Lambert Conformal Conic Projection, all rhumb lines will be straight lines.

1100. At 10,000 ft amsl and above, the reduced density of the atmosphere makes the use of an additional oxygen supply necessary to maintain efficient human metabolism.

1101. When flying from higher pressure to lower pressure the altimeter will over-read.

1102. The sequence of cirrus, cirrostratus, altostratus and then nimbostratus, heralds the passage of a cold front.

1103. (MAP) The adjacent symbol denotes an: Entry/Exit Point. ⊙

1104. During a Search and Rescue operation during the hours of daylight, the search aircraft will rock its wings if the ground signal(s) has been understood

1105. The International Datum Level is another name for the Standard setting of 1013.2 mb.

1106. The procedure to adopt in the event of an induction fire during engine starting will normally be found in the manual for the specific aircraft.

1107. Unless the pilot has prior permission from the ATCU an aircraft may not enter the aerodrome traffic zone in a Control Zone when the visibility is less than 5 nm or the cloud ceiling less than 3,000 ft.

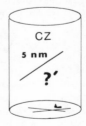

1108. All Military Air Traffic Zones have standard horizontal and vertical dimensions.

1109. Before undertaking a flight which will necessitate entering a Control Zone, the pilot should consult the RAC Section Vol. I of the UK Air Pilot to ascertain the exact procedure to follow in case of radio failure.

1110. Electrical fires in an aircraft cabin will normally break out without any warning symptoms.

1111. Granular, opaque rime ice may form on the wings of aircraft flying through clouds of super cooled water droplets.

1112. You are approaching from the east and you report airfield in sight. You are instructed: "G-RN JOIN DOWNWIND LEFT HAND CIRCUIT FOR RUNWAY THREE SIX". So you should overfly the airfield and turn left.

1113. The RPM of a fixed pitch propellor is control-led by the throttle setting only and is not affected by the speed of the aircraft.

1114. (MET) The letters GR mean Gale Report in force.

1115. The lift off speed will not vary with aircraft weight unless the aircraft is operating at more than the permitted maximum all up weight.

1116. The Aeronautical Emergency Service frequency 121.5 MHz should be used for all Distress communications.

1117. (MET) On a Significant Weather Chart the letters CB are used to indicate cumulonimbus.

1118.  A pilot who considers that his aircraft may have been endangered by another aircraft during flight should make an initial Airmiss Report within 7 days of the incident.

1119. RT frequencies used by ATC in specific Control Zones, Airways, Terminal Areas and those areas of FIR's which lie outside controlled airspace, are to be found in the AGA Section, Vol 2, of the UK Air Pilot.

1120. (MAP) The symbol signifies a 'Free Fall Parachuting' Area'.

1121. On the "Half Million" Topographical Chart" areas marked in green indicate open country suitable for emergency landings.

1122. (SIGNALS AREA) A red letter "L" or the shaft of a white dumbell signifies that light aircraft are permitted to take-off and land on either a runway or the area designated by a white letter "L".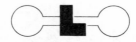

1123. A "Rate 2 Turn" is one in which the aircraft turns 360° in one minute.

1124. Wind speed is normally higher below 1,000 ft agl than it is above 2,000 ft agl.

1125. VOLMET is a broadcast made through the FIR service whenever any kind of hazardous weather conditions are reported.

1126. CAP 413 is a summary of the procedures used in RTF.

1127. 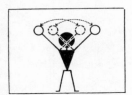 (MARSHALLING SIGNAL) Arms repeatedly crossed above the head, means: STOP.

1128. 'Wind Shear' is only produced in strong wind conditions.

1129. A kite is not an aircraft.

1130. Strong turbulence is more likely to be experienced during the passage of a cold front than during the passage of a warm front.

1131. Zone Control Units are sometimes established in Control Zones to control aircraft flying within the particular Zone, but not taking off or landing.

1132. (MAP) The adjacent symbol signifies the position of a military aerodrome with a runway or landing strip of 1829 m or over.

1133.  An aircraft shall not be flown in simulated instrument flight conditions without dual controls, a safety pilot and a competent observer to assist the safety pilot when necessary.

1134. When using a dry chemical fire extinguisher inside an aircraft cabin a pilot must anticipate a reduction in visibility and the posibility of difficulty in breathing.

1135. The SALR is approximately half the DALR.

1136. Induced drag accounts for less than 10% of the total drag at the low speeds employed during lift-off and landing.

1137. (MET) The adjacent symbol indicates: Heavy rain.   ●●

1138. There is more mixing between the lower level and higher level air by night than by day.

1139. An aeroplane which is banked needs to produce more lift than weight in order to maintain level flight. ·

1140. If a circuit breaker trips the pilot should re-set it immediately.

1141. The altitude of the Tropopause is higher at the poles than at the equator.

1142. (SIGNALS AREA) Two vertical yellow bars on a red square indicates that the landing area is serviceable but that normal safety facilities are not available. Aircraft may land in emergency only.

1143.

An aircraft which is flying in the United Kingdom in sight of the ground and following a road, railway, canal or coastline, or any other line of landmarks, shall keep such line of landmarks on its left.
Helicopters following certain landmarks may sometimes be exempt from this rule.

1144. With the altimeter set for QNE, an aeroplane cruising at FL 245 could be on a magnetic track between 270° and 359°.

1145. (MAP) The numerals shown above the adjacent symbol indicate height above mean sea level.

1950 ☀
(1720) 人

1146. A pilot is required to file a Flight Plan before any flight which crosses an International Boundary.

1147. The Transverse Mercator Projection is used for topographical charts covering small areas.

1148. As a general rule it is safe to fly beneath a thunderstorm if you can see through to the other side.

1149. At night, bursting projectiles showing red and green lights or stars, means that the aircraft is in danger of over-flying a prohibited area and should alter course immediately.

1150. Balance tabs move in the same direction as their elevators, and antibalance tabs move in the opposite direction to their stabilators.

1151. A single arrow on one side of a vector triangle of velocities indicates the side which represents HEADING and TAS.

1152.  The London and Scottish Flight Information Regions extend upwards to approximately 24,500 ft above which are their respective Upper Information Regions.

1153. If a fuse of lower amperage than that specified is fitted, the circuit can be overloaded resulting in a Fire Risk.

1154. (MAP) The symbol ↦ ↤ indicates: Cable joining obstructions.

1155. Whey flying over mountain areas the minimum safety altitude should be increased by at least 3000 ft.

1156. Whenever the fuel/air mixture has to be leaned, the mixture control should be ajusted to slightly on the rich side at max. RPM when below 3000 ft. a.m.s.l.

1157. In anticyclonic conditions mist and fog frequently occur in cold weather;; also the discending air often traps enough particles to cause a considerable haze.

1158. (MET) The adjacent symbol indicates: Severe wind shear in the lower levels.

1159. Vsₗ is the stalling speed, now power, flaps up and a load factor of I.

1160 When carrying out a Special VFR Flight over a built-up area, it is the pilot's responsibility under all circumstances to maintain a height of 1,200 feet above the highest fixed object within 2,000 feet of the aircraft.

1161. In temperate regions Nimbostratus will not form below approximately 12,000 feet.

1162. On Significant Weather Charts, scalloped lines are used to enclose areas of similar conditions, but low stratus and fog are not normally shown.

1163. Hill shading is not used to depict relief on the ICAO World Aeronautical Chart Series (1:1,000,000).

1164. When an aircraft banks in a turn towards the North or South the angle of dip in the Earth's magnetic field causes the magnetic compass to read incorrectly.

1165. 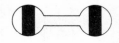 (SIGNALS AREA) A white dumbell with a black strip superimposed in each circular portion signifies that all movements of aeroplanes and gliders are confined to hard surfaces only.

1166. During a climbing turn the rate of climb will decrease unless power is increased.

1167. It is always illegal for a Student Pilot to operate an aircraft's radio without holding a "Flight Radiotelephony Operator's (Restricted) Licence".

1168. The carriage of aerosol cans and similar items packaged under pressure can present a hazard in light aircraft.

1169. Military Flight Training Areas have a radius of 5nm and extend from the surface to 3,000 feet aal, with a stub extending a further 5nm from 1,000 feet to 3,000 feet aal.

1170. A personal flying log book must be kept by persons applying for the grant or renewal of a licence, but it is not mandatory for members of flight crew to do so.

1171. When an aircraft is operated contrary to the conditions contained in the C of A, the pilot will, automatically, be fined £400.

United Kingdom Civil Aviation Authority
CERTIFICATE OF AIRWORTHINESS

1172. A flashing white light signal from aerodrome control to an aeroplane in flight means: LAND AT THIS AERODROME (CLEARANCES TO LAND AND TAXY WILL BE GIVEN AT THE APPROPRIATE TIME).

1173.  (MARSHALLING SIGNAL) Right arm down; left arm repeatedly moved upward and backward, means: TURN TO PORT.

1174. An upslope of 2° does not make an appreciable difference to the take-off run of a light aircraft.

1175. For a given control column movement a downward moving aileron rotates a small number of degrees than an upward moving aileron.

1176. The spoken word PAN on RTF indicates that the commander of the aircraft has an urgent message to transmit.

1177. A steady red light signal from aerodrome control to an aeroplane on the ground means: RETURN TO THE STARTING POINT ON THE AERODROME.

1178. Lines of Zero magnetic variation are known as "Isogonals".

1179. The term "wind velocity" refers to both the direction and the speed of the wind.

1180. The term "Hyperventilation" relates to a too fast breathing rate and can result in dizziness, nausea and blurred vision.

1181. The disposable load of an aircraft comprises the usable fuels and oils.

1182. It is not an offence to drop articles from an aeroplane in flight at a height of less than 100 feet agl.

1183. If the pitot tube becomes blocked by ice, the Airspeed Indicator will cease to function correctly.

1184. Aircraft which have demonstrated a satisfactory airworthiness standard, but which are not recognised by ICAO, are normally certificated in the Special Category or given a Permit to Fly.

1185. Water droplets or ice crystals which reduce visibility at the earth's surface to between 1,000 feet and 2,000 feet are called "mist".

1186. Information on airfield instrument approach charts published by the CAA is to be found in the CHARTS Section, Vol. 3 of the "UK Air Pilot".

1187. Every holder of a medical certificate issued under the Articles of the ANO should inform the CAA if he/she suffers any illness involving incapacity to undertake those functions for a period of 20 days or more.

1188. The vents in the tops of fuel tanks are for the constant equalization of pressure inside and outside the tanks.

1189. The word "Runway" is confined to paved areas provided for take-off and landing of aircraft.

1190. To obtain a Route Forecast in the form of a Significant Weather Chart, a pilot must give the appropriate meteorological unit at least 2 hours notice for routes of less than 500 nm and 4 hours notice for routes of more than 500 nm.

1191. Aeroplanes used in flying instruction for reward are issued with a Special Category C of A.

1192. Obstructions within 5 nm of an aerodrome boundary are listed in the RAC Section, Vol. 1, of the UK Air Pilot.

1193. (MET) The letters EMBD on a Significant Weather Chart mean: "Embedded".

1194. Even short exposure to high concentrations of carbon monoxide will seriously affect a pilot's ability to operate an aircraft.

1195. When a person is suffering from a head cold no ill effects will be experienced during flight.

1196. The magnetic variation over the UK is westerly.

1197. Lenticular clouds indicate low level convection currents.

1198. Nimbus clouds are rain-bearing clouds.

1199. A sensible rule in relation to alcohol and flying is to allow 8 hours between drinking and flying in order to allow for the effects of drinking large quantities.

1200. The mean chord line passes through the geometric centre of the wing from root to tip.

ANSWERS
X indicates "False"

1	41	81x	121x	161
2	42x	82	122x	162x
3x	43x	83x	123x	163
4x	44x	84x	124	164x
5x	45	85	125	165x
6	46x	86x	126x	166
7	47x	87x	127x	167
8	48	88x	128	168x
9x	49x	89	129	169
10x	50	90x	130x	170
11	51x	91	131x	171
12	52	92	132	172x
13x	53	93x	133x	173
14	54	94x	134x	174x
15	55	95	135	175x
16x	56x	96x	136x	176x
17x	57x	97	137	177
18	58	98x	138x	178x
19	59x	99	139	179x
20x	60	100	140	180
21	61	101x	141	181
22x	62x	102	142x	182
23x	63	103x	143x	183x
24	64x	104	144x	184x
25	65	105x	145	185
26x	66	106	146x	186
27x	67	107x	147	187x
28	68x	108	148x	188x
29	69	109x	149x	189
30x	70	110	150	190x
31x	71	111x	151	191
32	72x	112	152x	192
33x	73x	113x	153	193
34	74	114	154	194x
35x	75x	115	155	195x
36x	76x	116x	156x	196x
37	77	117x	157x	197
38	78	118	158	198x
39x	79x	119	159x	199
40	80	120	160x	200

201x	241x	281	321	361
202	242x	282	322x	362x
203x	243x	283x	323	363
204	244x	284x	324	364x
205x	245	285x	325x	365
206	246x	286	326x	366
207x	247	287x	327x	367x
208	248x	288x	328	368x
209x	249x	289	329	369
210	250	290	330x	370
211x	251x	291	331x	371
212x	252x	292x	332	372
213	253	293x	333	373x
214	254	294x	334x	374x
215	255	295	335x	375x
216x	256x	296x	336x	376
217x	257x	297	337	377
218	258	298	338	378
219	259x	299x	339	379x
220	260	300x	340x	380
221x	261	301x	341x	381
222x	262x	302	342x	382
223x	263	303x	343	383x
224	264x	304x	344	384x
225	265x	305x	345x	385x
226x	266	306	346	386
227x	267	307	347x	387
228	268x	308	348x	388x
229	269x	309x	349	389x
230x	270	310	350	390
231x	271	311x	351x	391
232	272x	312x	352	392
233	273	313	353x	393x
234x	274x	314	354	394x
235x	275x	315	355	395
236x	276	316x	356x	396x
237	277	317x	357x	397
238	278x	318	358	398
239	279x	319	359x	399x
240x	280	320x	360	400

401x	441x	481	521	561
402	442x	482x	522x	562x
403x	443x	483x	523x	563
404x	444	484	524	564x
405x	445x	485x	525	565
406	446	486	526x	566x
407	447x	487x	527x	567
408x	448x	488x	528	568x
409x	449	489	529	569
410	450	490	530x	570
411x	451x	491x	531x	571
412x	452x	492	532	572x
413	453	493x	533x	573x
414	454	494x	534x	574
415	455	495	535x	575x
416x	456x	496x	536	576x
417x	457x	497	537	577
418	458	498	538	578
419	459x	499	539x	579x
420x	460	500x	540	580x
421	461	501x	541	581
422x	462x	502	542x	582
423x	463	503	543x	583x
424	464x	504x	544	584x
425	465	505x	545x	585
426x	466	506x	546x	586x
427x	467	507	547x	587x
428	468x	508	548	588x
429	469	509x	549x	589
430x	470x	510	550	590x
431x	471	511x	551x	591
432	472x	512x	552x	592
433x	473x	513	553	593x
434x	474	514	554	594x
435x	475x	515	555	595
436	476x	516x	556x	596x
437x	477	517x	557x	597
438	478	518	558	598x
439	479x	519	559x	599
440	480	520x	560	600

601x	641x	681	721	761x
602x	642	682x	722	762
603	643	683	723	763x
604	644	684	724x	764
605	645x	685x	725x	765
606x	646	686	726	766x
607x	647	687	727	767x
608x	648x	688	728x	768
609	649	689x	729x	769x
610	650x	690	730	770x
611x	651	691x	731	771x
612x	652x	692x	732x	772
613	653x	693	733	773x
614x	654x	694	734	774
615x	655x	695x	735x	775
616	656	696	736	776
617	657	697x	737x	777x
618x	658x	698	738	778
619x	659	699x	739x	779
620	660x	700x	740x	780x
621x	661x	701	741x	781x
622	662	702x	742	782x
623	663x	703	743	783
624x	664	704x	744	784
625x	665x	705	745x	785x
626	666x	706x	746	786x
627	667x	707	747x	787
628x	668	708x	748	788
629x	669x	709	749	789x
630	670x	710x	750x	790
631	671x	711	751x	791x
632x	672	712x	752	792x
633	673	713	753x	793x
634x	674x	714x	754x	794
635	675	715x	755x	795
636	676	716	756	796
637x	677x	717	757	797x
638x	678x	718x	758x	798
639	679	719x	759	799x
640x	680x	720x	760	800x

801	841	881x	921x	961x
802x	842	882x	922	962
803	843	883	923x	963x
804x	844	884	924x	964
805	845x	885	925	965x
806x	846	886x	926	966x
807	847x	887	927	967
808x	848	888	928x	968
809	849	889x	929x	969x
810x	850x	890x	930	970x
811	851	891x	931	971x
812	852	892	932x	972x
813x	853x	893	933x	973
814x	854x	894	934	974
815x	855x	895x	935	975
816	856	896	936	976x
817	857	897x	937x	977x
818x	858x	898x	938x	978x
819x	859	899	939x	979
820x	860x	900	940	980x
821	861x	901	941	981x
822	862	902x	942	982x
823	863x	903	943x	983
824x	864	904	944x	984
825x	865	905	945	985
826	866x	906x	946x	986x
827	867x	907x	947	987x
828x	868	908x	948	988
829x	869	909	949x	989
830	870x	910x	950x	990x
831	871x	911	951	991x
832x	872	912	952x	992x
833x	873x	913x	953	993
834	874	914x	954x	994
835	875	915x	955x	995x
836	876x	916	956	996
837x	877x	917	957	997x
838x	878	918x	958x	998x
839x	879	919x	959	999
840	880x	920	960x	1000x

1001	1041	1081x	1121x	1161x
1002x	1042	1082	1122	1162
1003	1043	1083	1123	1163x
1004	1044x	1084x	1124x	1164
1005	1045	1085	1125x	1165x
1006x	1046x	1086x	1126	1166
1007x	1047	1087	1127	1167x
1008	1048	1088	1128x	1168
1009	1049x	1089x	1129x	1169x
1010x	1050x	1090x	1130	1170x
1011	1051	1091	1131	1171x
1012	1052	1092x	1132x	1172
1013x	1053x	1093	1133	1173
1014x	1054x	1094	1134	1174x
1015x	1055x	1095x	1135	1175
1016	1056	1096	1136x	1176
1017	1057	1097x	1137x	1177x
1018x	1058x	1098x	1138x	1178x
1019x	1059	1099x	1139	1179
1020	1060x	1100	1140x	1180
1021x	1061x	1101	1141x	1181x
1022	1062	1102x	1142	1182x
1023	1063x	1103x	1143	1183
1024x	1064	1104	1144x	1184
1025x	1065x	1105	1145	1185x
1026	1066x	1106	1146	1186
1027	1067x	1107x	1147	1187
1028x	1068	1108x	1148x	1188
1029x	1069x	1109	1149	1189x
1030	1070	1110x	1150x	1190
1031	1071x	1111	1151	1191x
1032x	1072	1112	1152	1192x
1033	1073	1113x	1153x	1193
1034	1074x	1114x	1154x	1194
1035	1075	1115x	1155x	1195x
1036x	1076	1116	1156	1196
1037	1077x	1117	1157	1197x
1038x	1078x	1118x	1158x	1198
1039x	1079	1119x	1159	1199x
1040x	1080x	1120	1160x	1200x